EVOLUTION IN
Changing
Environments

SOME THEORETICAL EXPLORATIONS

MONOGRAPHS IN POPULATION BIOLOGY

EDITED BY ROBERT H. MACARTHUR

EVOLUTION IN
Changing Environments

SOME THEORETICAL EXPLORATIONS

RICHARD LEVINS

PRINCETON, NEW JERSEY

PRINCETON UNIVERSITY PRESS

1968

FOR SARI

Preface

This book, except for Chapter 1, is based on a series of lectures, Frontiers of Population Biology, given in January 1965 to the Institute of Biology of the Cuban Academy of Science and in February of that year to the Biology Department at Yale University.

No attempt has been made at completeness. There is obviously an inordinate emphasis on my own work and that of my friends, since that is what I know best and what was most easily written up in isolation from good library facilities. As my contribution to the fight against the information explosion I have completely excluded those topics about which I have nothing to say regardless of their importance.

What remains is a series of explorations in fields where ecology, genetics, and evolutionary studies meet around the common theme of the consequences of environmental heterogeneity.

Many of the ideas presented here were developed in the course of collaboration or association with Robert MacArthur and Richard C. Lewontin, to both of whom I am greatly indebted.

I also give many thanks to Mrs. Lilian Altschul and to Miss Rita Luczynski, who typed the manuscript.

Preface to the Second Printing

Since the appearance of the first printing I have corresponded with a number of people who have pointed out errors or ambiguities, raised criticisms of some theoretical formulations, or have attempted to apply parts of the theory to particular systems. Meanwhile my own views have changed on some questions. I can only indicate briefly the directions in which a drastic rewrite would change the book:

1. *The metapopulation.* Any real species is a population of local populations which are established by colonists, survive for a while, send out migrants, and eventually disappear. The persistence of the species in a region depends on the rate of colonization successfully balancing the local extinction rate. But if every population has a finite life expectancy, the absolute distinction between stable and unstable communities breaks down into a continuum of extinction probabilities. These extinction probabilities increase as the smallest latent root of the community matrix of Chapter 3 decreases with increasing numbers of species. Therefore, the community matrix still can be used to predict the numbers of species, but in a slightly different way. Aspects of the metapopulation concept are under investigation in relation to group selection, biological control, and geographic distributions.

2. The measurement of competition presented in equations 3.9 and 3.10, although adequate for the Drosophila community, which provided most of the examples in Chapter 3, has hidden assumptions which are invalid for other situations. Competition is treated as if it is only the joint occurrence of individuals in the same microhabitat or food. Thus the P_{ih} sum to 1 for each species and intensities of interaction are ignored.

A new treatment begins with resources R_h entering the community at rate A_h and being harvested by the ith species at the rate $R_h P_{ih} X_i$. Then it can be shown that α_{ij} is found by the same equation as in Chapter 3 but with each $P_{ih} P_{jh}$

and each P_{ih}^2 weighted by the factor $\dfrac{R_h}{A_h}\, R_h$, the abundance
of the resource multiplied by the ratio of standing crop to
turnover rate. Further, P_{ih} is now an intensity measure as
well as a preference and need not sum to 1.

3. Some ecologists are unhappy about the whole notion
of the ecological niche because it is easily reified into a
fixed property of a species. In much of Chapter 3 the
niche is taken as given. However, since working with ants
I have been impressed with the great plasticity of the
niche, depending on which other species are present. The
theory of coexistence should examine the realized niche in
more detail, especially in relation to multidimensional
environments.

4. Although I recognized abstractly that the competition coefficients need not be constants, the consequences
of variable α's were not considered. One major class of
variable α's can no longer be ignored: much of the real
competition in nature occurs between the established populations of one species and the dispersal phase or immature
forms of another. The newcomer has a decided disadvantage,
and the outcome of competition often depends on priority.
This results in alternative equal stable communities and
patterns that must be studied at the metapopulation level.

5. The niche parameters are estimated in Chapter 3
and presented without confidence limits. I was hoping that
a better statistician than I am would respond to the
provocation and develop a sampling theory for the niche
parameters and community matrix. This is still pending.

6. Despite my claim that optima are not necessarily
achieved but merely indicate the direction of difference,
the whole tenor of the book is strongly selectionist. But
among students of biochemical evolution and polymorphism
at the protein level there is a growing conviction that much
of evolution is "non-Dominian" or random. These conflict-
ing approaches may be resolved along lines barely hinted
at in the last chapter: a large number of genes interact in
various ways to produce a smaller number of phenotypic
traits which are the direct objects of selection. Suppose that

for the ith trait there is some optimum function of the gene frequencies X_j:

$$f_i(x_1, x_2 \cdot \cdot \cdot x_n) = C_i.$$

For instance, if three genes interact additively in f_1 selection for f_1 puts gene frequencies on the plane

$$x_1 + x_2 + x_3 = C_1$$

the same genes may be recessives with regard to trait f_2, so that the optimum here is on the sphere

$$x_1{}^2 + x_2{}^2 + x_3{}^2 = C_2.$$

The joint action of selection for both aspects of fitness places the gene frequencies along the circle which is the intersection of the plane and the sphere where this lies in the interval $0 \leq x_j \leq 1$. Selection moves the population toward this locus, but movement along it is random. (The discussion in Chapter 3 considers what happens when the f_i have no simultaneous solution in the unit cube.)

Contents

EVOLUTION IN
Changing
Environments

SOME THEORETICAL EXPLORATIONS

CHAPTER ONE

On Theories and Models

Contemporary population biology has emerged in the last few years as a result of the convergence of the previously distinct disciplines of population genetics, population ecology, biogeography, and evolutionary studies in the context of a new holism and interest in theory.

The prevailing philosophy of American science has usually been reductionist and empiricist. And for a time the spectacular successes of molecular biology reinforced these attitudes. However, in the early sixties evolutionists and systematists began to defend the legitimacy of study at a population level. At the same time, questions of practical importance arose at the level of populations and environments—questions of pollution, conservation, biological control, environmental manipulation which involved the complexities of nature as essential ingredients.

Meanwhile, areas that were less amenable to reductionist analysis, such as developmental biology, also produced holistic currents. At a time when practical questions made the study of complex systems necessary, the independently arisen interest in systems research, cybernetics, and mathematical and theoretical biology created a favorable intellectual mood.

The relation of theory to experiment in biology has been an uneasy one. The word "theoretical" has generally had perjorative connotations, and the right to theorize was the reward for years of laboratory and field work. In fields where progress depends mostly on the refinement of technique in order to facilitate more accurate description, this perhaps did not matter too much, but other areas suffered from an indigestion of facts, while data was collected without reference to problems. In these circumstances, theoretical work often diverged too far from life and became exercises in mathematics inspired by biology rather than an analysis of living systems. This was accentuated also by the effort of theoreticians to model their work on physical systems.

Theoretical biology appeared on the stage as a correction to these biases. As against the prevailing empiricism it insisted on the right and need to theorize; as against growing specialization it insisted on an interdisciplinary approach; as against the dominant reductionism it emphasized the importance of complexity and holistic properties; as against the almost random lines of research in some fields it was biology self-conscious of its strategy. Therefore, theoretical biology as a distinct discipline is a transient phenomenon; once it has made its points part of the thinking of biology in general, it will loose its identity as a distinct field.

While practical interests created the necessity for an integrated population biology and a revived holism created a favorable climate, the internal logic of the component disciplines also forced this development. Increasing evidence has been accumulated to show that evolutionary, population genetic, biogeographic, and demographic events are not on entirely incommensurate time scales. The rate of evolution of say the size of the horse's foot may be 300,000 times slower than increase in bristle numbers in laboratory selection experiments. But Kurtén and others have shown that the slow rates of change on the scale of 50 million years may be the result of spurts of rapid change on the scale of tens of thousands of years, but with reversals of direction. Field observations have shown the remarkably rapid evolution of introduced species, or of species responding to changed environments over the last century. Experiments have shown that in the time it takes for species to interact demographically in simple competition, genetic changes can alter their competitive ability. Insular biogeography has made us aware of the extraordinarily rapid turnover of species, the rapidity of colonization, the high frequency of extinction. Natural selection has been observed in many cases to be strong enough to maintain differences between adjacent populations only meters apart in the face of high migration rates. Thus even the apparent stability of many biological situations can be regarded not as indicative of absence or weakness of evolutionary pressures, but rather as the dynamic balancing of strong forces, amenable to observation and measurement.

But the attempt to integrate fields that have developed independently leads to many difficulties. Some are difficulties in the translation of concepts from one area to another. Darwinian fitness (Wright's $\bar{W}$) has to be interpreted in terms of its ecological components, such as the intrinsic capacity for increase (Andrewartha and Birch's r_0) and the carrying capacity of the environment for a given genotype (K). Short-term fitnesses of this kind have to be related to the probability of long-term survival on a geological time scale. The ecologist's niche must be translatable to interpret the systematist's degree of specialization.

The analytical difficulties are also formidable. While the population geneticist's models generally assume stable age distributions, fixed population sizes, and constant environments in order to study the patterns of genetic heterogeneity in single species, the population ecologist considers genetically uniform populations in multispecies systems in a heterogeneous environment.

The attempt to consider genetic, demographic, environmental, and interspecific differences simultaneously immediately runs into technical difficulties. A precise mathematical description may involve hundreds of parameters, many of which are difficult to measure, and the solution of many simultaneous non-linear partial differential equations, which are usually insoluble, to get answers that are complicated expressions of the parameters which are uninterpretable.

We are clearly in need of a different methodology for coping with systems that are intrinsically complex. The following propositions define our strategy in approaching population biology; Numbers 1, 2, and 9 refer to our goals, while the others refer to methods. Thus they are propositions about science rather than about the objects of study. However, it is obvious where the methodological precept follows from a property of nature.

5

1. Given the essential heterogeneity within and among complex biological systems, our objective is not so much the discovery of universals as the accounting for differences. For example, instead of seeking a proposition of the form "there is a secular evolutionary tendency toward increasing complexity despite some exceptions," we would begin with the fact of different degrees of complexity and different directions of change, and ask what kinds of situations would give positive selective value to increased or decreased complexity. It may, of course, be argued that the result of such an inquiry would be a universal at a higher level. However, we would guess that such a law would also have a limited domain of relevance.

2. We are concerned more with the process of evolutionary and population dynamics than with their results, and are more interested in qualitative than in quantitative results. Except in applied problems which lie outside the scope of our present concern, numbers are of interest only insofar as they help in the testing of theory.

3. The basic unit of theoretical investigation is the model, which is a reconstruction of nature for the purpose of study. The legitimacy or illegitimacy of a particular reconstruction depends on the purpose of the study, and the detailed analysis of a model for purposes other than those for which it was constructed may be as meaningless as studying a map with a microscope.

4. A model is built by a process of abstraction which defines a set of sufficient parameters on the level of study, a process of simplification which is intended to leave intact the essential aspects of reality while removing distracting elements, and by the addition of patently unreal assumptions which are needed to facilitate study.

5. A sufficient parameter is an entity defined on a high level such as a population or a community which contains the combined relevant information of many parameters at a lower level. Thus genes act in many ways, affecting many physiological and morphological characteristics which are relevant to survival. All of these come together into the

sufficient parameter "fitness" or selective value. The way a gene contributes to fitness does not appear at all in the equations for genetic change. Similarly environmental fluctuation, patchiness, and productivity can be combined for some purposes in an over-all measure of environmental uncertainty, an important sufficient parameter. Since a sufficient parameter is a many-to-one transformation of lower-level phenomena, there is always a loss of information in going from one level to another.

6. There is no single, best all-purpose model. In particular, it is not possible to maximize simultaneously generality, realism, and precision. The models which are in current use among applied ecologists usually sacrifice generality for realism and precision; models proposed by those who enter biology by way of physics often sacrifice realism to generality and precision. The strategy which is followed in this book is the sacrifice of precision for generality and realism. This precision is sacrificed in several ways: we ignore supposedly unimportant factors each resulting in a small modification of the results; we ignore factors which will be important but only rarely; this means that there will necessarily be exceptions to all conclusions. Instead of specifying the exact form of a mathematical function we merely assume convexity or concavity, unimodality or bimodality, increasing or decreasing values. Therefore, the results come out as inequalities, and the hypotheses which are generated can only be tested by comparing whole groups of organisms.

7. The effects of the unrealistic "carrier" assumptions can be removed by replacing them with other sets of "carriers." A theorem which can be proved by means of different models having in common the aspects of reality under study but differing in the other details is called a robust theorem. Therefore the presentation of alternative proofs for the same result is not merely a mathematical exercise—it is a method of validation. Alternative models are also used as samplings from a space of possible models.

8. A theory is a cluster of models and their robust consequences. The constituent models fit together in several ways:

(a) as alternative schemes for testing robustness;
(b) as partially overlapping models which test the robustness of their common conclusions but also can give independent results;
(c) in a nested hierarchy, in which lower-level models account for the sufficient parameter taken as given on a higher level;
(d) as models differing in generality, realism, and precision (unlike the situation in formal mathematics, in science the general does not fully contain the particular as a special case. The loss of information in the process of ascending levels requires that auxiliary models be developed to return to the particular. Therefore, the "application" of a general model is not intellectually trivial, and the terms "higher" and "lower" refer not to the ranking of difficulties or of the scientists working at these levels but only levels of generality);
(e) as samples spanning the universe of possible models.

9. The role of general theoretical work is the following:

(a) The identification of the relevant sufficient parameters as new objects of study.
(b) The fairly direct generation of testable hypothesis.
(c) The posing of problems heading to lower-level theories which generate testable hypothesis. In this context, the general theory on adaptive strategy (Ch. 2) leads to the metahypothesis "strategic analysis of particular adaptive mechanisms will lead to verifiable hypotheses." The refutation of any particular one refutes only the lower-level theory, but if the bulk of the generated hypotheses are not confirmed the theory itself is invalidated.
(d) Besides prediction, a theory should offer plausible explanation for what we know. It is not necessary to derive the chemical properties of all macromolecules from quantum mechanical considerations, provided we can do so in a few cases and show that in principle it can be done for the rest. Similarly, it is no refutation

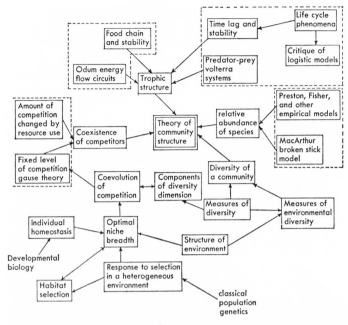

FIGURE 1.1. Relations among some of the components in evolutionary population biology theory. (From Levins, 1966.)

of evolutionary theory that we cannot at present account for the peculiar life cycle of migratory eels. The general theory argues that there is nothing unique to this trait which makes it impossible for it to arise by selection, and it could suggest several more specific models of the adaptive value of this trait.

Figure 1.1 shows schematically the cluster of models which defines the area of interest of these essays.

CHAPTER TWO

Strategies of Adaptation

The study of adaptation has usually meant the search for the adaptive significance of particular structures or physiological processes. This can sometimes be done directly in laboratory or field studies such as those on protective coloration. A trait can also be demonstrated to be adaptive by showing parallel variation in space or time for several species. In these studies, one assumes a constant environment and relates traits, for instance size to temperature, color to substrate, or hunting behavior to type of prey.

But there are other characteristics of organisms and populations which are not explicable as adaptations to particular environments—the degree of homeostasis in development, the amount of polymorphism, the extent of spatial differentiation of a species, sensitivity to natural selection, degree of inbreeding, and so on. These and other traits may be regarded as adaptations to the pattern of the environment in space and time, to temporal variability, to environmental uncertainty, and to what we will describe below as "grain." Such adaptations therefore fall into the category of strategies.

When our emphasis shifts to variable environments entirely new problems arise. If large body size is an adaptation to cold by way of the surface/volume relation, what size is optimum in an environment which is sometimes hot and sometimes cold? The commonsense of "folk liberalism" suggests a middle course, an intermediate size between those adapted to the extremes of heat and cold. But as we shall see this is only sometimes the case. In other circumstances a species may adapt completely to one of the extreme environments even at the expense of near lethality in the other. Or it may be polymorphic, containing a mixture of phenotypes (usually genetically determined), some of which are adapted to each environment. Or the animals may arrange to be large in the cold and small in the heat.

10

In the latter case, the environment may act directly on the developing system to evoke a phenotype which is more adapted to that environment. The response to the environment may be rapid and reversible, such as shivering in the cold, or it may be permanent, such as the size of holometabolous insects which is sensitive to larval temperature and is fixed throughout adult life.

There is always the danger that by the time the adaptation has taken place the environment has changed. This danger is of course greater when the delay is longer, and is especially so when the population adapts from generation to generation by genetic change through response to natural selection. Suppose for example that a species of butterfly has two generations per year, one in summer and one in winter. During the winter, those genotypes which are best adapted to winter conditions survive in greater numbers and leave more offspring to the next generation. Thus the next generation is better adapted to winter conditions than its parents but must develop during the summer. Similarly, in the summer generation those best adapted to summer conditions survive and produce a winter generation which is better adapted to summer. Such a species is always lagging behind the environment, doing the right thing for the previous situation. The result is to reduce fitness even more.

Thus we see that this species cannot adapt to the alternation of seasons by natural selection. However, each generation might adapt developmentally. The cold conditions faced by the larvae of the winter generation could produce cold-adapted adults which survive well in winter, and hot summer conditions may evoke the heat-adapted phenotype in the summer generation.

The same problem also arises in reversible physiological adaptation. Consider an inducible enzyme such as β-galactosidase which is formed to some bacteria in the presence of the sugar galactose. Is it better for the bacteria to contain the enzyme at all times in expectation that it may prove useful sometimes, or to produce the enzyme only in the presence of the substrate, or never to have the enzyme?

The induction of the enzyme takes time, roughly on the order of half an hour. Thus the cell which always has the enzyme gains about 30 minutes of growth whenever galactose appears as against those cells which must first synthesize it. However, it must tie up energy and protein in the form of an inactive enzyme when galactose is absent.

If the galactose is present or absent in bursts of 10 hours, the inducible cells lose 5% of the opportunity for growth, but pay only half the cost. Unless the cost is negligible it would seem that induction is the better strategy. But if galactose appears and is removed in periods of 30 minutes, the inducible cells still pay half the cost but now derive no benefit. In this case the permanent enzyme is the better strategy. Finally, if short spurts of galactose occur rarely enough it may be better never to utilize it.

This suggests that it would advantageous to speed the rate of response. If the enzyme could be fully synthesized in one minute, then half-hour fluctuations in the presence or absence of galactose could be met by induction. The environment at the time of induction would now be a good prediction of the environment after the response is completed.

Suppose, however, that galactose is introduced into the medium either in spurts of one minute or for ten minutes at a time. If the enzyme is induced after one minute in the presence of galactose, and if the short spurts are more frequent than the longer ones, the cost of synthesis may exceed the benefit derived from using galactose when it remains for ten minutes. Here perhaps a two-minute lag might be optimal. Now the cells lose 20% of the growth time in the long spurts but cut the cost in half by not responding to the short spurts. The optimal strategy must be able to distinguish between "signal" and "noise," doing so in this case by their different durations.

In the above cases the environmental factor to which the response is an adaptation is also the signal that evokes the response. But this need not be the case. Many insects go into a dormant diapause during the winter as an adaptation to the scarcity of food or to the cold. This diapause may be obligatory, in which case the insect is univoltine and has only one generation per year, or facultative, in which case the insect is multivoltine and only the last generation of the summer enters diapause. The determination of whether an animal will go into diapause or not often occurs in the egg. Although it may be an adaptation to the cold, the temperature over a few days is not a very good prediction of the temperature next month. The day length is generally the best single predictor, providing the best information as to the season. This brings us to the problem of environmental information in general.

The organism is conceived of as receiving this information through several receptor systems or through its whole body, processing this information into a prediction of future environments, and responding to this prediction with the formation of that phenotype which is the best strategy for the predicted range of environments. There is no necessary relation between the physical form of the information and the environmental factor to which the organism is responding. What is important is the statistical relation between them. Thus for most multivoltine insects a long day indicates summer and hence evokes direct development while a short day indicates autumn and evokes diapause. But in the silkworm moth, *Bombyx mora*, there are only two generations per year. The first is formed in early spring when the days are still short. It develops immediately in the short day and produces the second generation in summer when the days are longest. But here the long day predicts winter and brings on diapause.

The hypothesis that there is no necessary relation between the physical form of the signal and the response evoked is based on the assumption that there is a complex network of causal relations between the point of contact with the environment and the responding system. Different loops of this network can have opposite effects on the reacting system, some enhancing and some suppressing the original reaction. Therefore selection of an optimum response comes about by increasing some pathways and blocking others. The hypothesis is a useful one and suggests interesting experiments, but it will not hold universally. Where the

13

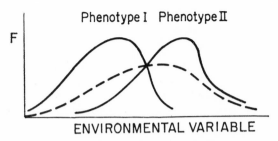

Phenotype I   Phenotype II

F

ENVIRONMENTAL VARIABLE

FIGURE 2.1a. Fitness as a function of the environment for two phenotypes. The dotted curve is the fitness of a mixture of the two phenotypes. The peak occurs at the optimal environment for each phenotype.

pathway between the environment and the responding system is short, its evolutionary flexibility will be much reduced. This seems to be the case for temperature. High temperature always accelerates development in the fully viable range and results in smaller body size in invertebrates, even where large size would seem advantageous. Some of the consequences of this will be discussed in Chapter 4.

We will now introduce the method of fitness sets for the analysis of adaptive strategy. In Figure 2.1a we show the relation between a component of fitness and the environment for an organism which may be in one of two physiological states. The location of the peak gives the optimal environment for that phenotype, the height of the peak is the measure of the best performance in the optimal environment, and the breadth of the curve is a measure of the tolerance for non-optimal environments. This tolerance is a measure of homeostasis. If there were no restrictions on the curve in Figure 2.1a, the optimal curve would obviously be infinitely high and infinitely broad. In fact there are restrictions. The height at the peak is undoubtedly limited by the physico-chemical structures involved. But in addition we suggest that the breadth cannot be increased without lowering the height. Suppose that the phenotypes I and II in the figure refer to two enzymes with different pH or temperature optima but with curves of the same shape. If the same total amount of enzyme is divided equally between the two forms the combined system has a fitness curve shown by the broken line in Figure 2.1a. If the enzymes are mixed in any proportions the shape of the curve may alter but the total area under the curve remains constant. If the curves refer to different genotypes in the same population the same argument holds.

The researches of George Sacher (1966) allow us to extend this principle of allocation further. It was first observed that despite differences in the total life spans of different animals, the caloric life spans measured in energy expenditure are remarkably uniform. The major discrepancies from constant expenditure were associated with the homeostatic system—the bigger the brain the lower the rate of aging per calorie. Thus Sacher defines organizational entropy as a measure of the energy cost per unit of development (carrying the organism from one stage to another). He amassed a great deal of data on his $S_{org}(T)$ as a function of temperature. It has a minimum value at some optimal temperature and increases with the departure of the temperature from the optimum. This "entropy" curve is broader and flatter for insect eggs and pupae than for larvae, which being mobile can seek out preferred environments. Sacher's work therefore suggests that the cybernetic system which reduces the organizational entropy is itself costly. It can be extended to cover a wider range of environments, but only by reducing the efficiency at the optimum.

Thus we assert the principle of allocation: the fitness curve $W(s)$ for an environmental parameter s may vary in shape but is subject to the constraint

$$\int F\{W(s)\}\ ds = C. \tag{2.1}$$

We do not know the functional form of F in general, but where fitness is altered by mixing components such as enzymes, $F(W) = W$.

For the purposes of this study we can specify that for any phenotype y and environment s, $W(s - y)$ is a non-negative function with a maximum at $s = y$ and decreasing symmetrically toward zero as $|s - y|$ increases. The dual of the curve in Figure 2.1a for a fitness component over environ-

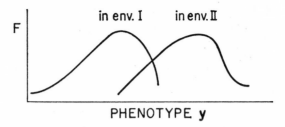

FIGURE 2.1b. Fitness as a function of phenotype in two environments. The peak occurs at the optimal phenotype for each environment.

ments is shown in Figure 2.1b, where for a fixed s we plot $W(s - y)$ over a range of phenotypes y. For any two environments the curves overlap. If they are close enough so that their inflection points overlap, the average of the two curves (fitness in an environment which is half S_1, half S_2) will have a single peak in the middle. If S_1 and S_2 are farther apart the average curve has a minimum at the midpoint and two peaks near S_1 and S_2.

The fitness set representation presents the curves of Figure 2.1b in a different way. The two axes in the graph are now fitness components W_1 and W_2 in environments S_1 and S_2. The phenotype whose peak is at S_1 in Figure 2.1b gives the point farthest to the right in Figure 2.2. The phenotype corresponding to S_2 gives the uppermost point. All pheno-

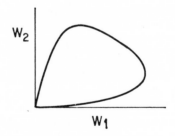

FIGURE 2.2. The fitness set. The coordinates of each point on the fitness set are the ordinates in Figure 2.1b, corresponding to the two environments for each phenotype.

types which have curves differing only in the location of their peaks will lie on the boundary of the fitness set shown in Figure 2.2. Those for which the area under the curve is less will lie inside the fitness set.

It can readily be seen that if S_1 and S_2 are sufficiently close (so that their inflection points overlap) the fitness set will be convex along the upper-right-hand boundary, while if S_1 and S_2 are farther apart the upper-right-hand boundary will be partly concave. This difference between the concavity and convexity of the fitness set will have important biological consequences. Therefore we define twice the distance from the peak to the inflection point as the tolerance of the phenotype and assert that the fitness set is convex or concave depending on whether the environmental range $|S_1 - S_2|$ is less than or greater than the tolerance of a single phenotype.

The importance of this distinction is that the fitness of a mixture of phenotypes in a population or of physiological states in an individual is represented by a point on the straight line joining their points on the fitness set. In a convex fitness set such mixed strategies will lie inside the set and therefore each one will be inferior to some single phenotype which lies above and to the right of it. But on a concave fitness set certain mixtures will lie outside (up and to the right) of the fitness set for single phenotypes so that mixed strategies may be optimal.

The fitness set alone does not define an optimum strategy. Over-all fitness in a heterogeneous environment depends on the fitnesses in the separate environments, but in a way which is determined by the pattern of environments. We therefore define the Adaptive Function $A(W_1, W_2)$, which measures fitness in the heterogeneous environment, to be a monotonic increasing function of its arguments. If the environment is sometimes S_1 and sometimes S_2, the individual must survive in both in order to survive. Let the probability of dying (or the loss in growth rate) in the interval Δt be $m(t)\, \Delta t$ where $m(t)$ takes on two values according to which environment S_1 or S_2 is currently present. Then the probability of survival up to time t is $P(t)$, and it satisfies the relation

$$P(t + \Delta t) = P(t)[1 - m(t)\,\Delta t]$$
$$\text{(assumes } m(t) \text{ unchanging for at least } \Delta t). \quad (2.2)$$

Where Δt is a whole generation,

$$P(t) = \Pi(1 - m(t)\,\Delta t) \qquad (2.3)$$

which is the product $W_1{}^p W_2{}^{1-p}$ for environments S_1 and S_2 occurring in the proportions $p : 1 - p$. But if Δt is very small the terms in $(\Delta t)^2$ and higher powers vanish, and

$$P(t) = P_0 e^{-\int m(t)\mathrm{d}t}. \qquad (2.4)$$

This will be maximized when the integral $\int m(t)\,\mathrm{d}t$ is smallest, which occurs when the linear average of the fitnesses $pW_1 + (1 - p)W_2$ is greatest. In the former case the environment is described as coarse-grained and presents itself to the individual as alternatives. Thus a coarse-grained environment is uncertain for the individual even if the proportions of the alternative environments remain fixed. In the latter case the environment is experienced as a succession of possibly different conditions. The differences present themselves to the organism as an average which is the same for all members of the population. There is no uncertainty. Of course there may be intermediate-grained environments, with the Adaptive Function intermediate between the linear and the multiplicative (or logarithmic).

The notion of grain comes of course from the size of patches of environment. If the patch is large enough so that the individual spends his whole life in a single patch, the grain is coarse, while if the patches are small enough so that the individual wanders among many patches the environment is fine-grained. But the concept can be made more general. For example, since most animals eat many times, food differences are fine-grained, whereas alternative hosts, when several are available for a parasite, are coarse-grained differences.

Thus for the situations we have described the Adaptive Function $A(W_1, W_2)$ may vary from the hyperbola-like $W_1{}^p W_2{}^{1-p}$ to the linear $pW_1 + (1 - p)W_2$. In any case the

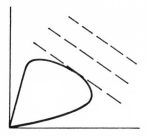

FIGURE 2.3a. Fine-grained environment on a convex fitness set.

optimal strategy is represented by the point on the fitness set (single phenotype or mixed strategy) which touches the curve $A(W_1,W_2) = K$ at the greatest K. In Figures 2.3a, 2.3b, and 2.3c we show several optimum strategies. In formal terms the results are the following:

1. On a convex fitness set (environmental range smaller than the tolerance) the optimum strategy is a single phenotype which is adapted to some intermediate value of S between S_1 and S_2, and does moderately well in both these environments.

2. On a concave fitness set (environmental range exceeds the tolerance) a fine-grained environment results in an optimum strategy of a single phenotype which is specialized to either environment S_1 or S_2 depending on p, the frequency of environment I. Therefore it does optimally in one environment and poorly in the other.

3. On a concave fitness set with a coarse-grained environment the optimum is a mixed strategy in which the two

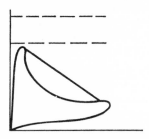

FIGURE 2.3b. Fine-grained environment on a concave fitness set.

19

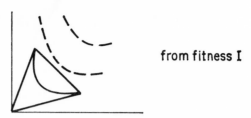

from fitness I

FIGURE 2.3c. Coarse-grained environment on a concave fitness set.

specialized phenotypes occur in proportions that depend on p.

When there are more than two environments new axes have to be added, and in a continuous environment we need infinitely many dimensions for the fitness set. However, as S_1 becomes arbitrarily close to S_2 any phenotype which does well in one will do well in the other, and the fitness set becomes a straight-line segment at 45° from the origin. Thus we only have to consider as distinct those environments different enough to change the ranking of fitnesses of the available phenotypes.

An alternative approach considers the environmentally determined optimum phenotype S to have a probability distribution $P(S)$, and attempts to maximize $\int A[W(S)]P(S)\,dS$ subject to the restriction $\int W(S)\,dS = C$. When $A(W)$ is linear, the optimum allocation is specialization to the most common S, while for a coarse-grained environment $A(W)$ is $W_1^p W_2^{1-p}$ or equivalently $p \log W_1 + (1 - p) \log W_2$, and the optimum is $W(S) = CP(S)$. Thus the fitness is assigned to each environment in proportion to its frequency. Qualitatively the results are the same—coarse-grained environments are uncertain and give rise to less specialized broad-niched populations, with the niche breadth increased by the uncertainty of the environment. The niche breadth interpretation is pursued further in Chapter 3.

On the basis of this formal theory we can now interpret biological situations by defining "phenotype" in various ways.

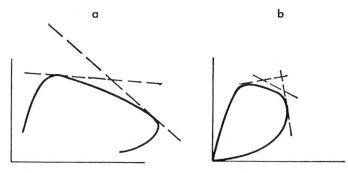

FIGURE 2.4a. Optimum phenotype as a function of the probability of a particular environment p_e. A small change in p_e alters the slope of the Adaptive Function, moving the optimum phenotype a long distance on a flat fitness set.

FIGURE 2.4b. Optimum phenotype as a function of the probability of a particular environment p_e. Here a small change in the position of a point on the fitness set corresponds to a large change of slope.

A developing organism may respond to a variable environment in several ways:

1. It may always produce the same phenotype, regardless of the environment. This corresponds to Schmalhausen's (1949) autonomous development and Waddington's (1957) canalization. This would be the optimum strategy in the absence of environmental information—when fluctuations are quite rapid compared to the generation or when there is no signal that is predictive of future environments and which the organism can process effectively. This stability of a given phenotype may of course depend on some internal strategies of other kinds which minimize the effects of environmental perturbations.

2. It may produce a phenotype which varies as a continuous function of the environmental signal (Schmalhausen's dependent development). In Figure 2.4 we show how the optimum phenotype varies with the probability of environment I, p_e. The subscript indicates that p_e is a conditional probability, the probability of environment I given the environmental signal e. But the phenotype changes at a rate which depends on the shape of the fitness set. If the

fitness set is more or less flat, a small change in p_e will move the point of tangency a great distance along the fitness set, which means that a large change of phenotype is involved. But for a very convex fitness set as shown in Figure 2.4b almost all slopes p_e can be found along a short segment of the fitness set, so that in the range near $p_e = 0.5$ there is only slight change in the optimum phenotype with the environment. The sensitivity of the optimal phenotype to the environment also depends on the grain. In a fine-grained environment the slope of the Adaptive Function is

$$\partial W_1/\partial W_2 = -(1 - p)/p \qquad (2.5)$$

so that the derivative of the slope with respect to p_e is

$$\partial(\partial W_1/\partial W_2)/\partial p = 1/p^2. \qquad (2.6)$$

For a coarse-grained environment the corresponding derivative is

$$\partial(\partial W_1/\partial W_2)/\partial p = (1/p^2)W_1/W_2. \qquad (2.7)$$

Since p is the proportion of environment I, when p is large W_1/W_2 is also large. As p decreases, $1/p^2$ increases but W_1/W_2 decreases, so that a given change in p produces a smaller change in the optimum phenotype.

3. It may produce one phenotype for environments below a given threshold and a second phenotype above that threshold (Schmalhausen's autonomous regulative development, or developmental switch). This is the optimum strategy for a fine-grained environment when the range of environment exceeds the tolerance of the individual phenotype.

4. It may produce a mixture of phenotypes, the probability of each depending on the environment. This type of development, a stochastic switch, would be optimal for coarse-grained environments when the range exceeds the individual tolerance.

Patterns 1 and 2 are the most familiar ones. While they are understandable in a general way, the quantitative interpretation of rates and degrees of acclimatization is not obvious. We know for example that temperature acclimation requires somewhat less than one day for amphibians (Brattstrom and Lawrence, 1962) and for *Drosophila* (our work), but substantially longer for fish. Also, different species acclimate to different degrees. Among the Puerto Rican *Drosophila*, the narrow-niched species were all poor at acclimation to heat. But the broad-niched species included two which acclimate well, one which had a very high heat resistance without acclimation, and one which hardly acclimated at all.

The switch patterns are also widespread. They include the determination of sexual vs. asexual eggs in the Rotifera, dormancy in many seed plants, diapause in insects and mites, the production of winged vs. wingless aphids, formation of free-living vs. infective larvae in some nematodes including Strongyloides, aerial vs. aquatic leaves in some plants, etc.

E. O. Wilson (unpublished) used the dual form of the fitness set to analyze caste polymorphism in ants. We transpose his results here into our fitness set notation. Wilson considers that an ant colony faces many crises of different kinds that can reduce the production of queens, which is the ultimate measure of fitness. Different phenotypes can cope differentially with these situations. For any two situations the fitness set may be represented as in Figure 2.5. When the fitness set is convex, the optimum is a single all-purpose caste. But for a concave fitness set (which means, for very different situations) polymorphism may be optimal. Wilson postulates that selection pressure will be greatest on those traits involved in meeting the most serious or most frequent contingencies, so that finally for all contingencies i with frequency p_i and fitness contribution W_i, $p_i W_i =$ constant. Thus his Adaptive Function maximizes the total fitness $\Sigma p_i W_i$ subject to the restriction $p_i W_i = p_j W_j$. Thus the optimum population is composed of that mixture of casts whose fitness point lies on the line $p_1 W_1 = p_2 W_2$ where it crosses the boundary of the fitness set. Unspecialized castes are roughly equally fit for both contingencies and therefore are represented by points near the 45° line, while specialized

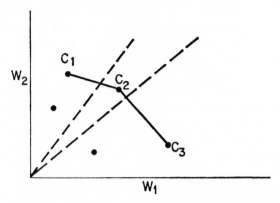

FIGURE 2.5a. A change in the relative weights assigned to different contingencies may result in the replacement of one caste by another.

castes lie closer to the two axes. From these considerations the following can be proved:

1. The number of castes cannot exceed the number of kinds of contingencies to be met. This follows from the fact that the line $p_1 W_1 = p_2 W_2$ always passes between a pair of

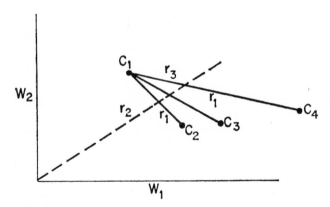

FIGURE 2.5b. The optimal population consists of two castes in the proportions of the line segments divided by the Adaptive Function (dotted line). Thus castes C_1 and C_2 are more or less equally common, but a population of C_1 and the more specialized C_3 will be mostly C_1.

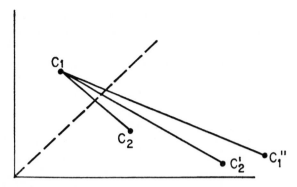

FIGURE 2.5c. Selection favors increased specialization. As C_2 increases in specialization its fitness point moves to the right. The mixed populations form a new fitness set which is superior to the previous one.

fitness points. (We exclude the infinitely unlikely event that three fitness points are colinear.)

2. A change in p will produce greater changes in caste proportions if these are unspecialized (near the 45° line). A pair of specialized castes is a more stable arrangement, less likely to be broken down during evolution, than a pair of unspecialized castes. But as Figure 2.5a shows, in a polymorphism of one unspecialized and one specialized caste the unspecialized caste is retained as specialized castes replace each other.

3. The more efficient a caste at meeting the contingency for which it is specialized, the lower its abundance. This is seen in Figure 2.5b.

4. Increased specialization of a caste, improving its fitness at one task, usually increases colony fitness even when it reduces the effectiveness of that caste for other tasks. This is shown in Figure 2.5c.

Strategic analysis has been applied by Martin Cody (1966) to the problem of clutch size in birds. He uses a three-dimensional fitness set whose axes are clutch size, predator avoidance, and competitive ability. He assumes a convex fitness set so that it is not necessary to specify the Adaptive Function at all except that it increases along each axis. Since clutch size would be an aspect of r-selection (selection for the rate of population increase when it is below saturation), it will have greater weight in environments which are unstable from generation to generation, and populations may often be below saturation. This leads to the prediction that clutch size will increase not only from the tropics to the temperate zone (which is already known) but also from the coast inland and from sea level to higher elevations. Further, tropical regions with great variation in resources such as are found for sea birds off the coast of Peru will also show large clutch size compared to close relatives elsewhere.

Toward regions of greater environmental stability the rate of predation is usually greater, so that the maximization of the Adaptive Function would result in decreasing clutch size but an increase in the anti-predation expenditure. However, for birds which are well protected by nesting in holes, and for many island dwelling birds, predation is much reduced, and the tropics-temperate gradient in clutch size will therefore be less steep.

Cody considered a single fitness set for all birds. A more detailed study of the biology of different families would show predictable differences in the convexity of the fitness set, which would affect the steepness of the gradient. In fact, the case of the hole-nesting species could be looked at in this way. The choice of a nest site in a hole does not reduce the ability to feed. Therefore, when predator avoidance is based on hole nesting instead of on making fewer visits to the nest, the fitness set is more convex on the predator-avoidance–clutch-size plane. Therefore a given change in the slope of the Adaptive Function would result in a smaller change in clutch size. Instead, Cody allowed hole nesting to reduce the weight of predator avoidance in the Adaptive Function. Both methods are equally valid. The division of a biological problem into the fitness set and the Adaptive Function is to some extent arbitrary. We have found that it is usually most convenient to divide the difficulties of the problem more or less equally between the two.

The problem of delayed reproduction is a widespread one, including delayed maturity, seed dormancy, resting spores in some invertebrates, and diapause. We can treat the diapause problem as representative of these questions.

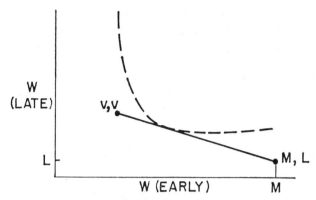

FIGURE 2.6. A fitness set analysis of diapause strategy. The point v,v represents diapausing forms, while M,L represents the non-diapausing phenotype.

Let the axes of the fitness set be fitness in early summer (when there is time for another generation) and in late summer (when there is no longer time). The phenotypes are duration of diapause, from zero to several months. The intermediate-length diapause reduces fitness in both environments—reproduction is lower due to the delay but the breaking of diapause in midwinter is also lethal. Therefore the fitness set is concave, and we only have to consider the two phenotypes complete versus no diapause.

The dormant eggs have a small probability of death, which for convenience we assume to be about the same in early and late summer. The non-dormant eggs are virtually inviable in late summer but will leave many offspring if it is early summer. Thus the fitness set is highly asymmetric. This is shown in Figure 2.6.

The alternative environments are "coarse-grained" so that the Adaptive Function is $p \log W_1 + (1 - p) \log W_2$. It has a slope of $-(1 - p)W_1/pW_2$, where p is the probability of environment I. The relevant part of the fitness set is the line joining the points for diapause and non-diapause. If v is the viability of dormant eggs, M is the viability of non-dormant eggs in early summer, and L is the viability of non-dormant eggs in late summer, the fitness set has a slope of $-(M - v)/(v - L)$. The optimum strategy may be all eggs developing directly, all in diapause, or some fraction of the eggs in each class. Let the proportion of eggs developing directly without diapause be q. Then the rate of increase if good conditions prevail is $\log \{v + q(M - v)\}$ and if winter comes within a generation, fitness is $\log \{v - q(v - L)\}$. Thus the average fitness is $p \log\{v + q(M - v)\} + (1 - p) \log \{v - q(v - L)\}$.

Hence:

$$\hat{q} = v \left(\frac{p}{v - L} - \frac{(1 - p)}{M - v} \right)$$

provided that this lies between 0 and 1. Hence:

$$\hat{q} = \begin{cases} 0 & p \le \dfrac{v - L}{M - L} \\[2ex] v \left(\dfrac{p}{v - L} - \dfrac{(1 - p)}{M - v} \right) & \text{for } \dfrac{v - L}{M - L} \le p \le \dfrac{M}{v} \left(\dfrac{v - L}{M - L} \right) \\[2ex] 1 & p > \dfrac{M}{v} \left(\dfrac{v - L}{M - L} \right) \end{cases}$$

$$(2.8)$$

Therefore for $M = 1$, $L = .1$, $v = .9$, the optimum q is 0 when p is less than $.8/.9$ or about .89 and rises to 1 when $p = .8/.9 \times 1/.9$ or about .97. For a large value of M the transition from no diapause to complete diapause is less abrupt. For $M = 10$ and the other terms unchanged, $\hat{q} = 0$ when p is below $.8/9.9 = .08$, and rises to 1 at $p = .89$. Thus the same kind of adaptive system can result in the relatively abrupt switch from 0 to 100% diapause, or can produce a slower gradual transition as in the production of winged vs. wingless aphids.

Dan Cohen (1967b) has approached the problem of seed dormancy analytically and numerically. Among his conclusions is the observation that an optimizing system would not generally utilize all the available environmental information. Signals that would evoke the same response can be lumped, and a new source of information will be used only if it would lead to different responses which increase fitness sufficiently to offset the cost of reception. In another study the same author (1967) analyzes the sufficient signal for bird migration.

Monte Lloyd (1966) considered a very unusual kind of delayed reproduction—the 13-year and 17-year generations in the periodic cicadas. While the origin and maintenance of this pattern is difficult to account for, the adaptive significance seems to be the escape from control by predators. A predator whose life cycle is less than the generation length of the cicada would be out of phase with the cicada years most of the time, since 13 and 17 are prime numbers.

Strategic analysis can also be applied to behavior. Here we consider only one aspect of behavior: active habitat selection. Consider a species in which the individual has a limited time in which to find a suitable microhabitat. Let there be two kinds of usable sites, a preferred one with viability M and a less satisfactory site where fitness is L. The density of habitat type I is p, so that the probability of finding one such site in time t is $1 - e^{-\lambda p t}$, where λ is a measure of searching ability or mobility and t is the available time. The density of habitat II is $q = 1 - p$. However, if this habitat is found it will only be used with a probability θ. The problem is to optimize θ.

Total fitness is the fitness if a site is found times the probability of finding a site. This is

$$W = \left[\frac{pM}{p + \theta q} + \frac{\theta q L}{p + \theta q} \right] [1 - e^{-\lambda(p+\theta q)t}]. \quad (2.9)$$

The first bracket is equal to $M - (M - L)\theta q/(p + \theta q)$ and therefore decreases with θ. The second term increases with θ. Clearly, if $M - L$ is negligible the optimum $\theta = 1$, while if λt is large enough the second term will be near 1 regardless of θ, and the optimum θ would be zero.

In a heterogeneous population, $M - L$ depends on the gene frequency. For each gene frequency there is a value of $M - L$, and this determines the optimum θ. But θ, by determining the effective proportions of the two environments which the population actually faces, can determine gene frequency. Thus there is a coevolution of specialization and habitat preference, which will be examined in Chapter 3. We see that for a fixed $M - L$, the degree of habitat specialization is determined by $1 - e^{-\lambda t}$, the probability of finding a usable site. This is a measure of the effective productivity of the environment for the species in question. Low productivity is in effect uncertainty and, like other kinds of uncertainty, results in a broadening of the niche (a reduction of specialization).

MacArthur and Pianka (1966) have applied the same kind of argument to hunting behavior. They distinguish between searching and pursuing. During the search the hunter is not committed to any particular kind of prey (except indirectly by selecting the habitat where it searches). In contrast, pursuit implies the chasing of a particular food item and therefore precludes hunting for anything else. MacArthur and Pianka reason that if the search time is long compared to the pursuit time, very little search opportunity is lost during pursuit. It is therefore advantageous to pursue any capturable item. But if pursuit is long compared to search the energy invested in pursuit might be better invested in searching for the preferred items.

Hence pursuers may specialize more readily in their choice of food than searchers can. The relative importance of searching and pursuing is partly a property of the biology of the animal and partly a property of the abundance of food. Once again high productivity leads to low uncertainty and specialization.

We now pass to a consideration of population strategies based on genetic make up and the response to natural selection. Here, in addition to the principle of allocation, there are constraints imposed by the genetic system. Some of these will be discussed in Chapter 5. Here we note only the following:

1. If the fitness set is convex, the optimum strategy is monomorphic. But the optimum phenotype may vary with a slowly changing environment. The population can either remain monomorphic at that phenotype which is the aver-

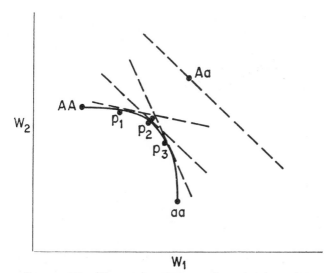

FIGURE 2.7a. Fitness set with genetic constraints. Average heterosis permits polymorphism in a fine-grained environment provided both environments are abundant enough (the Adaptive Function not too horizontal or vertical).

age optimum, or change in response to selection. But the response to selection requires some genetic variance within the population, thus preventing complete monomorphism.

2. A stable polymorphism of specified optimal types may segregate non-optimal types as well.

Both of these departures from optimality are part of the cost of maintaining a strategy.

Consider first a habitat with two environments. Let a population show variation for a phenotype controlled by a single locus with two alleles. The fitnesses of the three possible genotypes can be shown on the fitness set in Figure 2.7. Any panmictic population can be represented by a point on the curve (AA,aa). This curve bisects the median from the heterozygous vertex Aa to the base AA,aa of the triangle in the figure. Inbreeding has the effect of flattening the curve toward the base AA,aa.

In a fine-grained environment the Adaptive Function is $pW_1 + (1 - p)W_2$. Therefore (Figures 2.7a,b) polymorphism will only be optimal when there is average hetero-

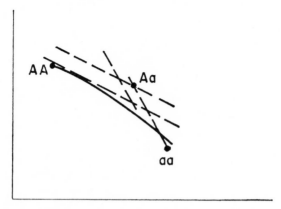

FIGURE 2.7b. With a flatter fitness set the range of polymorphism is restricted.

sis (that is, Aa is above and to the right of the line joining
the homozygous points). Further, if p is too close to 0 or 1
the population will be homozygous. Unlike the previous
fitness-set arguments, an environment which is too rare
will have no effect on the optimum.

Where there is polymorphism in this model, it is not an
optimum mixed strategy in the sense of the previous discussion. The optimum would still be monomorphic, of the
heterozygous phenotype. But a population of all heterozygotes is not normally possible, so that the polymorphism
is imposed by segregation and in fact reduces fitness. In
Figure 2.7c we see the segregation load.

With fine-grained environment and a concave fitness set,
one or the other allele is fixed. There is therefore no con-
tinuing response to selection for different environmental
frequencies, whereas on a convex fitness set each frequency
of environments determines a gene frequency. The sensitivity of gene frequency to the environment depends on the
shape of the fitness set. In Figure 2.7a we see a very convex
set. As we argued previously, a change in the slope of the
Adaptive Function moves the equilibrium point (which is
also the optimum strategy subject to the constraints of the
genetic situation) along the fitness set to a point with the
same slope. For a highly convex set, this is accomplished

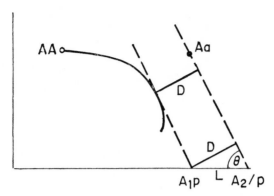

FIGURE 2.7c. The fitness loss due to segregation pL. $D/L = \cos\theta$, and $\tan\theta = -p/(1 - p)$.

along a relatively short arc. Thus even wide fluctuations in the environment will produce only small changes in gene frequency. This polymorphism will be highly stable. Such a situation is of course advantageous when there is little correlation between the environments of successive generations so that the optimum gene frequency is hardly altered. But note that this stability is achieved at the cost of a large segregation load.

In Figure 2.7b the fitness set is much flatter. Here the loss of fitness due to segregation is less, but the gene frequency varies more as the environment changes. The fitness set can be flattened by reducing the fitness of the heterozygote or by lengthening the line AA,aa. This is equivalent to increasing the difference between homozygotes.

In a coarse-grained environment the situation is different. It has been discussed in Levins and MacArthur (1966) and will be examined in subsequent chapters.

A more precise description of the behavior of a genetic system under selection can be gotten from the model in which there is an optimum phenotype S which is an environmentally controlled random variable. For each phenotype y, fitness declines with the square deviation from optimum so that

$$W = 1 - (S - y)^2/H. \qquad (2.10)$$

TABLE 2.1. Genetic model for selection in a varying environment

Genotype	Frequency	Phenotype	Fitness
AA	x^2	a	$1 - (s - a)^2/H$
AA'	$2x(1 - x)$	0	$1 - s^2/H$
A'A'	$(1 - x)^2$	$-a$	$1 - (s + a)^2/H$

Here H is a measure of homeostasis. The model for a single locus with two alleles and no dominance is shown in Table 2.1. The parameter a is the average effect of the allele A on the phenotype. The fitness of the population at any one time is

$$\bar{W} = 1 - (S - M)^2/H - VAPHE/H, \qquad (2.11)$$

where M is the mean phenotype of the population at that time and $VAPHE$ is the genetic variance at that time. The average fitness over time is then seen to be

$$E(\bar{W}) = 1 - (1/H)(\sigma_S^2 + \sigma_M^2 + E(VAPHE) - 2\,\mathrm{cov}\,(S,M)). \qquad (2.12)$$

Thus fitness is reduced by environmental variance, by genetic change (σ_M^2), and by the average variance in the population. The first component is imposed from without, but the other two are part of the cost of strategy. This fitness loss may be offset by the covariance term provided the correlation between the environments of the successive generations is high enough (for this model, about 0.8).

It was shown (Levins, 1965) that the amount of genetic variance needed to carry out the optimum tracking of the environment is quite small, at least an order of magnitude less than the variance of the environment. Further, the optimum strategy is not very efficient. It only restored up to one-third of the fitness lost by environmental fluctuation. Of course other models might give more efficient tracking.

The approach outlined here places environmental heterogeneity at the center of our analysis. A number of environmental properties are identified as sufficient parameters: the environmental range, the uncertainty, the grain, temporal variance, autocorrelation between environments of successive periods, and cross correlations between environmental factors which allow signals such as photoperiod to act as predictors of physically unrelated factors. We also specified several properties of the organism which are taken as given: the fitness set, which gives the relation between fitnesses in different environments for each potential phenotype; the information-receiving capacity of the system; and in some cases the genetics. These organismic properties interact with the environmental ones to give the optimal strategies. These are monomorphic specialization, monomorphic generalization (intermediate in both environments), fixed polymorphism (mixed strategy), and environmental tracking.

The optimum strategy may be found at each of several levels. Once established, it determines the fitness set for the next higher level. Thus a mixed strategy in the sense of a mixture of isoenzymes results in a broader tolerance of the synthetic process involved than can be achieved by one enzyme type alone. This would make the fitness set for cellular types more convex. Thus adaptations occur in an interlocking hierarchy at various levels within the individual physiology, through behavior, and at the level of population structure and dynamics. But at each level the same general principles emerge:

1. Environmental heterogeneity reduces fitness below what it would be in a uniform environment.

2. An optimum strategy can be found subject to the restrictions imposed by the biological situation. This optimum strategy reduces but cannot eliminate the fitness loss due to environmental heterogeneity.

3. The optimum strategy itself imposes a cost on the system. Therefore it is not enough that an adaptive mechanism works—it must increase fitness in terms of the original phenotype-environment system by more than some threshold amount (the cost) before it will be advantageous.

4. At all levels, increased environmental uncertainty results in optimum strategies which spread fitness over a

broader range of environments. If the environmental range exceeds the capacity of a single adaptive system, a mixed strategy will be optimum at that level.

There are a number of major gaps in the theory. For instance, we have been assuming all along a single environmental component and a single corresponding phenotype. But of course most adaptive characters are relevant to survival in several ways, and each component of fitness is affected by many aspects of phenotype. Formally we could assert that environment and phenotype are both vectors rather than scalars, but that would obscure a number of possibilities that we will not explore further now but merely list:

1. A mixed strategy may involve different parts of the phenotype. For example, a plant which is surviving in rain forest and also in a drier forest may have the large rounded leaves typical of the first environment but drop them seasonally as befits plants of the deciduous forest. Or it may have the opposite, with evergreen habit but a more xeric leaf form. There are many possible combinations which would result in great morphological variety even in the same kind of strategy.

2. This creates new possibilities for shifting a trait from one adaptive context to another. For instance, in the anoline lizards pigmentation is relevant to temperature relations and to camouflage. If a dark pigmentation allows an animal to warm up quickly in a cool habitat, and also matches the color there, the same trait is doing double duty and there is no problem. But if the habitat is cool and the substrate is light then what is helpful with respect to temperature is harmful for predator avoidance. Unpublished observations by Harold Heatwole suggest that the conflict is partly resolved by behavior. He found that the better-camouflaged lizards could be approached quite closely before they fled, but those with poor camouflage fled when the herpetologist was farther away. Insofar as flight reactions take over the burden of predator evasion, the body color is released to evolve with respect to temperature.

36

3. When an organism is adapted to a complete environment, and the different components begin to separate with climatic change, what strategy would be followed? Suppose for instance that a species is at its optimum temperature and substrate, but that secular change raises the temperature. It can move toward the new location of its optimum temperature, remain in its optimum substrate, select an intermediate habitat which is moderately satisfactory for both, or perhaps migrate so that the crucial stages of the life cycle have optimum temperature and substrate.

4. Since each part of the organism is the environment for other parts, we can study the evolution of internal diversity as the coevolution of parts and processes. Some examples of this are considered in the last chapter.

The theory outlined here suggests several lines of experiment and observation. First, it leads to direct predictions of a general sort expressed as inequalities comparing whole groups, such as the temperate and tropical faunae, tree and hole-nesting birds, or ant species with generalized and specialized caste polymorphisms.

A second line involves the detailed analysis of particular adaptive systems with the measurement of the fitness set, environmental uncertainty and information, and the actual strategy compared with the calculated optimum strategy. Diapause strategy in economically important insects or mites would be especially amenable to such study since there is widespread monitoring of their populations by agricultural experiment stations. We have already described a rough qualitative model. Further refinements for particular cases would lead to specific predictions as to the date of onset of diapause at different locations, population differences in the importance of different signals, and the uniformity of the switch from non-diapause to diapause forms. It would also lead to predictions as to the results of attempting to control the pests by confounding their information. If artificial lighting were used to delay the onset of diapause, light would cease to be a good predictor of season and the populations affected might become obligate diapausing univoltine races.

A third line of work is in the genetics of flexibility for different types of structures and processes. If the pattern of response is strategic, it would be selected separately for different aspects of the same organism instead of being controlled by some general "flexibility" factors, and may vary widely in the same character from group to group. The work of Bradshaw (1965) and his collaborators seems to support this view.

Finally, one could attempt to produce natural selection for an optimal strategy in laboratory populations. Since "strategy" is a phenotypic trait, there is no reason to doubt the possibility of such selection. The differentiation of the European corn borer (*Ostrinia nubilalis*) populations into distinct diapause races in the forty or so years since their introduction into North America encourages the belief that the selection may be strong enough to be rapidly effective. One such experiment would be to subject *Drosophila* larvae to a single day of high or low temperature in the last instar. The high-temperature treatment would be followed by selection of the first flies to emerge from the puparium. The low-temperature treatment would be followed by discarding the first flies and retaining the last group only. We would therefore expect natural selection for a pupal dormancy or pseudodiapause evoked by one day of later larval temperature. The next step from there would be to replace the temperature signal by light, or oil of wintergreen. We could then test the hypothesis of the arbitrary physical relation between the form of the signal and the evoked response, that almost any signal can be coupled to almost any responding system.

Beyond the specific experimental and observational testing of the theory, this approach directs attention toward the sufficient parameters of the environment, and would therefore direct bioclimatological work toward the definition and measurement of these evolutionarily significant characteristics of environment.

Finally, strategic analysis can be used to interpret the evolution of complex systems such as biochemical networks and behavioral modes, suggesting relevant measures and selection pressures.

CHAPTER THREE

The Theory of the Niche

The concept of the ecological niche has been used heuristically for a long time. Grinnell (1904) referred to the niches of birds to indicate that different species have different requirements. Elton (1927) used the niche mostly for defining a species' position in the trophic hierarchy. Biogeographers have long noted the apparent equivalence between organisms of different regions and corresponding numbers of species in corresponding formations. For instance, in Table 3.1 we show the percentages of bird species of the Maylayan and Australian rain forests in each level of the forest and for each food category. This correspondence suggests that the relative abundance of opportunities for species of each type, or the number niches, is the same in both forests. The notion of saturated environments versus those with empty niches has been used to compare insular and continental biotas. Finally, the classical experiments of Gause (1934) led to the exclusion principal: if two species coexist they must occupy different niches. But it was due to the work and inspiration of G. E. Hutchinson (1965) that progress has been made toward a rigorous theory of the niche.

A satisfactory theory of the niche must permit an accurate description of a species' niche, and must be able to use that description to answer the following questions:

1. What determines the degree of specialization of a species, or inversely, its niche breadth?
2. What determines the species diversity of a community in relation to area, climatic region, size of organism, trophic level, etc?
3. How similar can species be and yet coexist?
4. How do similar species divide the environment among them?

Table 3.1. The subdivision of the rain forest habitat among bird species (from Harrison 1962)

	Per cent of 306 Maylayan birds (per cent of 117 Australian birds)				
	Herbi-vore	Carni-vore	Insecti-vore	Omni-vore	Total
Air	0(0)	2(7)	4(5)	0(0)	6(12)
Upper canopy	13(9)	0(0)	5(1)	8(11)	26(21)
Middle zone	0(3)	4(5)	40(42)	10(7)	53(57)
Terrestrial	1(2)	0(0)	8(4)	6(4)	15(10)
Total	14(14)	6(12)	57(52)	24(22)	100(100)

5. How do species in the same community affect each others' evolution? When do species alter their niches?

There are a number of ways of representing the niche of a species in an abstract hyperplane. One method would be to let each axis represent an environmental factor such as temperature, size of food particle, etc. Then each point in the space represents a set of environmental components, and if the species can survive (or is found) in that environment the point is included in the niche. The niche is then a region of the environment space. This definition may be refined somewhat to indicate how well a species must do in a given environment for it to be included in the niche. In this representation, niche breadth appears as the area of the niche in the hyperplane.

Another representation increases the number of axes. Instead of a single temperature axis there are axes for minimum and maximum temperature (or for midpoint and range). Then the niche is reduced to a point. Here niches can no longer overlap, but similarity appears inversely as the geometric distance between points.

We prefer to define the niche as a fitness measure on an environment space. Figure 3.1 shows the temperature niche for two species of *Drosophila*.

We have found that the following measures are a set of sufficient parameters for the theory of the niche and community:

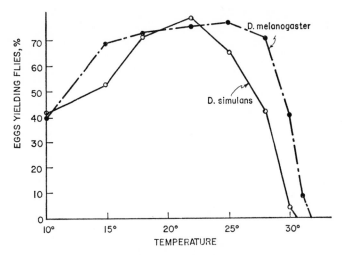

FIGURE 3.1. The temperature/emergence relation for two species of *Drosophila*. Clearly *D. melanogaster* has a broader temperature niche than *D. simulans*. (After Tantawy and Mallah, 1961.)

1. niche breadth;
2. niche dimension;
3. niche overlap;
4. community diversity.

NICHE BREADTH

Data on niche breadth come from three sources:

1. Survival experiments such as those of Tantawy's in Figure 3.1. Since fitness requires not only survival but also successful reproduction, this is clearly not a complete fitness measure but is an important component. In the figure it is seen that *D. simulans* has a narrower, higher, more specialized temperature niche than *D. melanogaster*. Any measure of spread could be used to quantify niche breadth.

2. Habitat or food selectivity in multiple choice experiments. For example, Martinez et al. (1965) set out several different kinds of bait (banana, tomato, potato, and oranges) in *Drosophila* traps less than ten feet apart. Thus any fly caught on any bait could have reached any of the others.

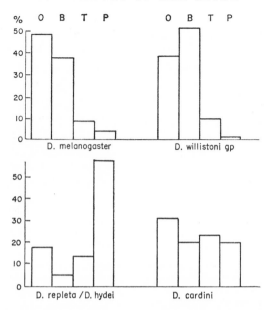

FIGURE 3.2. Food preference histograms for four species of *Drosophila*. O, orange; B, banana; T, tomato; P, potato. (From Martinez Pico et al., 1965.)

Some of their results are shown in Figure 3.2. A species such as *D. cardini*, which is attracted to each bait with almost equal frequency, would be said to have a broad niche for food as compared to *D. repleta*. Of course the baits offered did not span the whole range of *Drosophila* food, but it did include fruits differing in acidity, starchiness, and sugar content. All the flies were drawn from the same population, and since we did not observe evidence of aggressive exclusion of flies, the distributions are assumed to indicate true preferences and hence to correspond to the pre-competitive or potential niches.

3. Actual distributions of species over environments. (a) Environmental factor identifiable. We can use the frequency distribution of tidal organisms over levels across a beach, plant species vs. altitude, or flies against season of the year. (b) The environmental factor not identifiable but known to vary. Then the uniformity of a distribution over a presumed

patchy environment indicates a broad niche, and extreme clustering suggests a narrow niche provided we can exclude a clustering tendency per se and the persistence of progeny at the site of their birth. When the patches are small enough compared to the mobility of the species, the potential niche is measured.

Maldonado and Levins (in preparation) studied the microhabitat niche of *Drosophila* by setting out 20 traps with the same banana bait in a small area and classifying the species taken in each trap separately. Some of their results are shown in Table 3.2.

TABLE 3.2. Microhabitat niche breadths for *Drosophila*. $B = 1/\Sigma p_{ij}^{2}$

	Collection 5	Collection 6
Maximum	14	13
D. melanogaster	2.4	3.5
D. willistoni	5.3	7.5
D. latifasciaeformis	7.0	8.5
D. Dunni	5.4	5.2
D. ananassae	5.4	4.0
All flies	6.1	8.5

Two measures of niche breadth have been proposed:

$$\log B = -\Sigma p_i \log p_i \qquad (3.1)$$

and

$$1/B = \Sigma p_i^2. \qquad (3.2)$$

In both measures, p_i is the proportion of the species which is found in environment i, which selects environment i, or in the case of a viability measure

$$p_i = v_i \Big/ \sum_i v_i \qquad (3.3)$$

where v_i is the viability in environment i.

There is no very strong reason to prefer one measure over the other as yet. Both give niche breadths equal to N for N equally used resources or for uniform utilization over an interval of length N and no utilization outside. And both measures are similar, as is shown in Table 3.3. Finally, since the number of environmental classes is arbitrary, the meas-

TABLE 3.3. Two measures of seasonal niche breadth for *Drosophila*.*
Measure $1 = 1/\Sigma p_{ij}^2$; measure $2 = \exp(-p_{ij} \log p_{ij})$; maximum
value is 25. Data for Austin, Texas, from Patterson, 1943

Species	B1	B2
All	11.14	14.27
D. melanogaster + simulans	9.58	12.06
D. hydei	3.64	6.24
D. mulleri + aldrichi	4.23	6.92
D. macrospina	8.30	10.71
D. longicorni	4.39	6.95
D. affinis + algonquin	9.81	12.58
D. hematofila	3.51	5.16
D. putrida	3.52	6.16
D. pseudoobscura	4.65	6.51
D. melanica	10.35	13.07
D. busckii	3.21	4.51
D. meridiana	3.06	3.69
D. immigrans	1.70	1.84
D. robusta	2.53	3.44
D. tripunctata	6.72	9.35
Average	4.08	5.35
Correlation	.98	

* Species with fewer than 100 flies omitted.

ure B should be divided by the maximum number, which is
the number of classes, to give comparable measures of niche
breadth.

Once we have a measure of niche breadth we can ask
whether the abundant species tend to have broader niches
than the rare ones, whether climax species have narrower
niches than colonizing species. We can also compare niche
breadths for the biotae of different regions, zones, trophic
levels, or taxonomic groups. For the Puerto Rican *Drosophila*
we have found that the abundant species are usually the
ones which are broad-niched.

Immediately the question arises, how can we tell if we
are measuring the relevant factors so that the calculated
niche breadths have real meaning. We will show below that
the niche description leads to predictions of numbers of
species and other community properties which enable us to
check its completeness.

In the previous chapter we reached several conclusions about niche breadth. Qualitatively they all suggested that a broad niche is optimal in an environment which is uncertain. This uncertainty may derive from temporal variation in the environment from generation to generation, from a coarse-grained patchiness which is uncertain for each individual, or from a low density of usable resources or habitats (low productivity of the environment for the species in question). In a stable environment, fitness will be spread out only over environments which are so similar as to give a convex fitness set.

One difficulty with the theory is that while the species is allowed to adapt to a pattern of environmental heterogeneity that pattern is taken as given. Yet we know that habitat or food preferences may reduce the effective environmental heterogeneity. The amount of niche reduction by behavioral preferences depends on the viability niche breadth and on the productivity. We showed in Chapter 2 that fitness, $\bar{W}$, was given by

$$\bar{W} = [M - (M - L)\theta q/(p + \theta q)][1 - \exp(-\lambda(p + \theta q)t)],$$
(3.4)

where M is the fitness in the better habitat, L is fitness in the poorer habitat, p and q are the relative frequencies of the two habitats (or resources), θ (between zero and one) is the probability of accepting the less favored environment, λ is a measure of the probability of encountering a unit of either habitat per unit time, and t is the available time. We saw that for a fixed p,λ,t the optimum value of θ may be zero when $M - L$ is very large and the optimal θ is one when $M = L$ (or when $M - L$ is small enough). But for a population $M - L$ depends on the gene frequency. In Table 3.4 we show a model of a one-locus, two-allele system with symmetric effects. This is also shown in Figure 3.3. We took $h > 0$ in order to have a convex fitness set.

As seen from the figure, the equilibrium gene frequency will be zero if p is small enough and one if p is close enough to one. For the population as a whole, the difference in fitness between the two environments is $M - L$ when the

TABLE 3.4. Model of fitness in two environments

Genotype	Frequency	Fitness in environment I	II
AA	x^2	M	L
AA'	$2x(1 - x)$	$(M + L)/2 + h$	$(M + L)/2 + h$
A'A'	$(1 - x)^2$	L	M

gene frequency of A is zero, $-(M - L)$ when the frequency is one, and zero at some intermediate gene frequency (here at $1/2$, due to the symmetry).

In Figures 3.4a,b we represent this fitness difference D of the equilibrium population as a function of $p^* = p/(p + \theta q)$, the effective frequency of environment I.

Since D is the result of natural selection, we claim that for each p^*, D approaches the solid curve. In terms of gene frequency x, $D = (2x - 1)(M - L)$.

But for each D there is some optimum θ which determines an optimum p^*. For $D = 0$, the optimum θ is one and $p^* = p$. For some D sufficiently large, the optimum θ falls below one and can decrease to zero. Thus p^* will evolve toward the broken line in Figures 3.3a,b. The joint evolution of D and p^* (that is, of viability niche breadth and habitat

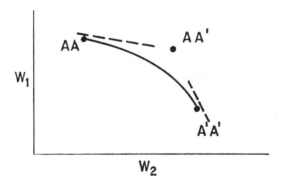

FIGURE 3.3. Model for selection in two environments. The equilibrium population is the point of tangency of the fitness set (solid curve) with the Adaptive Function $A = pW_1 + (1 - p)W_2$ (the broken lines).

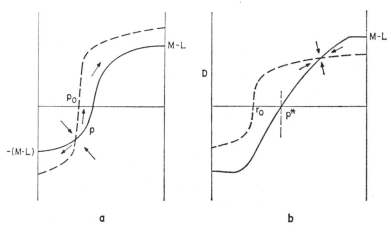

FIGURE 3.4a and b. The coevolution of habitat preference and niche breadth. D (the difference between fitness in environments I and II) evolves toward the solid line, while p^*, the effective frequency of habitat I, is a consequence of behavior and evolves toward the broken line. p_0 is the proportion of habitat I in the environment. In 3.4a the result is a stable equilibrium, in 3.4b there are two alternative specialized modes of adaptation.

selection) is shown by the arrows in the figures. If productivity is sufficiently low even the maximum D (at $x = 0,1$) may be insufficient to favor habitat selectivity. Then there is a stable equilibrium at $p^* = p$ and D less than $M - L$. But if productivity is high enough so that habitat specialization is favored at $D = \mp (M - L)$, there is an unstable equilibrium and two stable states corresponding to full specialization to the two alternative environments. Finally, asymmetry in the model may result in an unstable equilibrium separating one specialized and one unspecialized equilibrium. Here past history as well as the ecological situation determines the outcome.

NICHE DIMENSION

If species divide their habitat among themselves on the basis of a single factor such as temperature, their niches can be represented along a single dimension as in Figure 3.5.

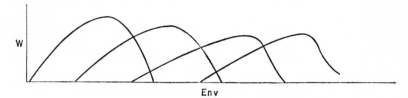

FIGURE 3.5. A one-dimensional array of niches.

This has the consequence that each species competes primarily with its two neighbors and hardly at all with the others. If two factors are distinguished, as in Harrison's classification of rain forest birds, competition is spread more evenly among several species which overlap in different directions. Thus niche dimensionality refers not to the number of biologically relevant factors in the environment, which may be virtually unlimited, but to the number of factors which serve to separate species.

No formal conventions have yet been adopted for measuring niche dimension. However we can already ask how many niche dimensions there are in a community and their relative importance.

MacArthur (1958) has analyzed the bird environment in terms of vegetation profile. He found that the relative density of vegetation at three levels corresponding to the herbaceous and low shrub, middle, and upper levels of the forest is sufficient to predict the occurrence of warbler species within the deciduous forest of the northeastern United States.

Our experiments with *Drosophila* were mostly limited to the species attracted to fruit bait on the ground, thus excluding the pollen and fungus feeders. We found differences among species in food preference, microhabitat within the forest, sites within a region, time of day, and season. We approached the dimensionality by way of a measure of community diversity

$$D = -\Sigma f_i \log f_i, \tag{3.5}$$

where f_i is the relative abundance of species i. When several

TABLE 3.5. Contributions of niche components to diversity. Diversity measures given as logarithms to base 10

	Texas*	Puerto Rico	Brazil†	Belém (Brazil)
Within trap		.28		
Among microhabitats		.25		
Total with collection	.53	.53	.67	.56
Seasonal	.22	.16	.12	.03
Total within site	.75	.69	.79	.59
Among sites in zone ⎱	.25	.22		
Among zones ⎰		.10		
Total	1.00	.91		

Note: The among zones and sites figure for "Texas" comes from the whole southwestern United States.
* From Patterson (1943).
† From Dobzhansky and Pavan (1950).

collections are made the total community diversity is equal to the average diversity within collections plus the diversity due to differences among collections. Therefore by using different seasons, baits, or sites as "collections" we can measure the relative importance of niche components along different dimensions. The food component is an overestimate, since it was based on setting out traps of different kinds of bait in equal numbers. In nature the non-uniform distribution of foods would reduce their contribution to diversity. The diurnal component of diversity does not directly separate species, since flies can come to the same bait at different hours. However it is an indirect indication of different microhabitat preferences. The niche components for our Puerto Rican *Drosophila* are shown in Table 3.5 along with comparable data from Brazil and the United States. We note that total diversity is roughly comparable for all regions but that the seasonal component of diversity falls from 30% of the total at Austin, Texas, to 22% in Puerto Rico, to 16% for all Brazilian collections, and to 3% for Belém. This gives a more meaningful measure of environmental variability for flies than would the meteorological data alone.

The relatively small contribution of geographic sites is striking. This is perhaps a result peculiar to small animals. Within a forest there is considerable variation in temperature and humidity from one cubic centimeter to another. Almost all the range of environments that can be found among forests at all elevations in Puerto Rico can occur somewhere within the same forest. Similarly, the diversity found at Austin alone is only increased by one-third when we include the whole southwest. This would of course not be true for species with very narrow food preferences such as *Drosophila carcinophila*, which breeds on the gills of land crabs in the Caribbean, or *D. peninsularis*, which is mostly a littoral species in areas with cactus.

We expect the situation to be different with regard to larger animals. The variation from one quarter acre to another is not as great as that from cubic centimeter to cubic centimeter, and will therefore include a smaller part of the total variation over a geographic range. For the trees of the forest, zonation is also quite prominent.

Thus we expect the comparative study of niche dimensions to be an important part of community ecology.

NICHE OVERLAP

At issue here is the amount of competition or of ecological similarity among species. One measure would be some geometric distance, such as

$$d_{i,j} = \sum_h (p_{ih} - p_{jh})^2 \tag{3.6}$$

where p_{ih} is the niche measure of species i in environment h. However not all biological differences reduce competition. Different birds may capture the same insects in flight or on a tree; fruit flies may use the same fruit at different hours. The p_{ih} must be limited to components of competition. In a coarse-grained environment, where individuals spend their whole lives in the same patch, viability in each patch type separately affects the competition, but in a fine-grained environment this will not be true.

The above measure has been used (Martinez et al., 1965)

TABLE 3.6. Maximum number of species for which the expected value of the community is positive

Average α	Covariance (α_{ij} with α_{ji})									
	.005	.01	.015	.02	.025	.03	.05	.075	.1	.3
.1	34	27	19	15	13	11	8	6	5	3
.2	24	22	16	13	11	10	7	6	4	
.3	21	18	13	11	9	8	6	5	4	
.4	17	14	11	9	7	7	5	4	4	
.5	14	11	8	7	6	5	4	3	3	
.6	12	8	—	5	5	5	3	3	3	

for food differences. However we prefer to measure niche overlap by the coefficient α_{ij} used by Gause. He started from the Volterra equations for increase of species x and y:

$$dx/dt = rx(K - x - \alpha y)/K \qquad (3.7)$$

and

$$dy/dt = ry(K - \beta x - y)/K. \qquad (3.8)$$

The coefficients α and β measure the reduction in the rate of increase of x caused by an individual of y compared to the effect of an individual of x, and vice versa. Since this competition depends on the probability that individuals of the two species meet (in the sense of seeking food in the same habitat, or pursuing the same kind of prey) a good approximation for α_{ij} is

$$\alpha_{ij} = \sum_h p_{ih} p_{jh} \Big/ \sum_h p_{ih}^2 \qquad (3.9)$$

which is

$$\alpha_{ij} = \sum_h p_{ih} p_{jh}(B_i), \qquad (3.10)$$

where B_i is the niche breadth. Thus α_{ij} is similar to a regression coefficient of one species' use of environment on that of the other. We see that because of the factor B_i, α_{ij} will not equal α_{ji} unless their niche breadths are equal. There is the unexpected result that α_{ij} can exceed one even without special mechanisms, such as environmental poisoning. A broad-niched species spread out over many environments will have a lower rate of encounter between members than would a more specialized species. Suppose for example that

TABLE 3.7. Coefficients of competition between species. First number is microhabitat α; number in parentheses is seasonal α. From unpublished studies at Mayagüez, Puerto Rico

Species	mel	lat	will	Dun	ana
D. melanogaster	1(1)	.30(.61)	.42(.76)	.61(.55)	.16(.75)
D. latifasciaeformis	.72(.70)	1(1)	.92(.79)	.72(.48)	.60(.50)
D. willistoni gp.	.88(.95)	.81(.85)	1(1)	.96(.57)	.47(.59)
D. Dunni	.90(.34)	.44(.26)	.67(.29)	1(1)	.38(.63)
D. ananassae	.18(.60)	.28(.35)	.25(.38)	.29(.81)	1(1)

one species uses environments A and B with frequencies $2/3$ and $1/3$, and that the second species is limited to the first environment. The probability of encounters between members of the first species is $(2/3)^2 + (1/3)^2$ or $5/9$, the encounters of the second species with itself have a relative frequency of 1, and the encounters between species occur only in habitat A at the rate of $1 \times 2/3 = 6/9$. Therefore $\alpha_{12} = 6/5$ and $\alpha_{21} = 2/3$. This does not mean however that species 2 will exclude species 1, because the broad-niched species may have a greater K.

The coefficient of competition α can be measured with respect to any aspect of the environment separately, and the over-all α will then be a product of the individual ones. Once we have defined α we can raise questions about its statistical distribution. What is the average level of competition (how closely are niches packed), how variable is α, how does this differ in young and mature communities, etc.? In Table 3.7 we show some α's obtained from Puerto Rican *Drosophila*.

The logistic equations 3.7, 3.8 have been criticized from a number of points of view. However the use of the bracketed term in the equation to define the equilibrium conditions does not depend on the validity of the equations for describing the rate of change toward equilibrium. Nor does it matter that the α_{ij} may vary with population density.

THE COMMUNITY MATRIX

The simultaneous equations

$$dx_i/dt = r_i x_i (K_i - x_i - \Sigma \alpha_{ij} x_j)/K_i \qquad (3.11)$$

can give an equilibrium community when

$$K_i = x_i + \Sigma\alpha_{ij}x_j \tag{3.12}$$

for all i. These equations can be expressed as a single matrix equation

$$AX = K, \tag{3.12}$$

where X is the column vector $\begin{vmatrix} x_1 \\ x_2 \\ x_3 \\ \cdot \\ \cdot \\ \cdot \end{vmatrix}$,

K is the vector of the K_i, and A is the community matrix

$$A = \begin{vmatrix} 1 & \alpha_{12} & \alpha_{13} & \cdot\ \cdot\ \cdot \\ \alpha_{21} & 1 & \alpha_{23} & \cdot\ \cdot\ \cdot \\ \alpha_{31} & \alpha_{32} & 1 & \cdot\ \cdot\ \cdot \\ \cdot & & & \quad 1 \end{vmatrix}$$

whose elements α_{ij} are the competition coefficients. Even without the solving of equations the matrix gives much information about the community. For competitors the α's are positive, while for predator-prey pairs α_{ij} and α_{ji} have opposite sign. For competitive α's, the average value indicates how closely species are packed. If the niche is one-dimensional, each species can have only a few positive α's and the others will be near zero. As the dimensionality of the niche increases the variance of the α's can be reduced. Thus the variance of the α's is an indicator of dimensionality. For competitors, $\alpha_{ij} = \alpha_{ji}$ when the niches are equal in breadth. Then the pairs of symmetrically arranged terms have a correlation of $+1$. As the niche breadths become more variable, this correlation decreases, and it is also reduced by predator-prey pairs.

With the descriptive parameters of the niche defined and the community matrix described, we can use them to study the original questions.

The simultaneous equations 3.12 could be solved to get the relative abundances if we knew the α_{ij} and the K_i. We have suggested how to measure the α_{ij}, but K_i is less directly observable, since it is the maximum population attainable in the absence of competitors. Therefore we reversed the procedure, and used the known x_i and the calculated α's to find K_i. This was done for microhabitat separation in our Puerto Rican *Drosophila*. In Table 3.8 we show the frequencies of

TABLE 3.8. Carrying capacity K and relative niche breadth $B/\bar{B}$. From unpublished data, collections at Mayagüez, Puerto Rico, 12/26/65. $\bar{B}$ is the breadth for all flies combined.

Species	Fre-quency	K	$B/\bar{B}$	$K/(B/\bar{B})$
D. melanogaster	.03	.41	.42	.98
D. willistoni group	.54	.89	.87	1.02
D. latifasciaeformis	.32	.92	1.00	.92
D. Dunni	.01	.62	.61	1.02
D. ananassae	.08	.27	.47	.62
D. tripunctata species	.02	.35	.44	.79

species, their niche breadths B relative to the breadth for all flies, $\bar{B}$, the estimated K, and the ratio of K to $B/\bar{B}$. The uniformity of this ratio suggests the hypothesis for more general testing of whether the carrying capacity K is proportional to the niche breadth B.

There is no *a priori* reason to expect this result. It means that the species do not differ very much in the average efficiency with which they utilize their resources but only in how wide a range of microhabitats they can use. Another surprising result is the suggestion that in spite of seasonal changes in relative abundance of the species, at each moment the community is in equilibrium for the current environment but this equilibrium changes with the changes in the resources, keeping the community close to a moving equilibrium. Thus the community matrix approach helps to determine if we have in fact identified the important environmental factors and to determine if the community is near equilibrium.

At a more abstract level the community matrix leads to some results about the limit to the number of coexisting species. In order for a community to be stable the determinant of its matrix must be positive. Furthermore, the symmetrized determinant whose element $\alpha_{ij}^* = \alpha_{ji}^* = (\alpha_{ij} + \alpha_{ji})/2$ must be positive, and so must each subdeterminant formed by striking out a row and the corresponding column. Therefore we can investigate the average value of these determinants in terms of the statistical distribution of the α. In Appendix I (at the end of this chapter) we derive the recurrence relation for the expected value D_n of a determinant of rank n (n rows and columns). The result is the pair of equations

$$D_n = D_{n-1} - (n-1)\bar{\alpha}^2 T_{n-1} - (n-1) \text{ cov } (\alpha_{ij},\alpha_{ji})D_{n-2}$$
$$(3.13)$$
$$T_n = D_{n-1} - (n-1)\bar{\alpha}T_{n-1} \qquad (3.14)$$

where the initial values $D_0, D_1 = 1$. We see that the covariance term enters in such a way as to reduce D_n. Thus a community in which niches are equally broad can hold fewer species than one with high dimension (uniform α's, low variance) and non-uniform niche breadths. We suspect that as a community matures the variance of the α's decreases and more species could be accomodated. Thus a waif fauna of diverse origins should reach a demographic equilibrium with fewer species than old faunae hold. Recent work by Wilson and Taylor on Pacific ants (1967) seems to support this idea.

In the next section we will use the equations 3.13, 3.14 to estimate the number of species which can coexist in a community given their general similarity and heterogeneity.

THE NUMBER OF SPECIES IN A COMMUNITY

Gause's analysis of competition for a part of species lead to the conclusion that for equal K's, any $\alpha < 1$ permits coexistence. This was interpreted to mean that the two species must use the environment in different ways and therefore that there must be at least two distinguishable resources to support two species.

This can be extended to the more general theorem: The number of species cannot exceed the number of distinct resources. In Appendix II (at the end of this chapter) we offer two proofs of this. The first uses the simultaneous equations (3.12) and our definition of α_{ij}. It therefore assumes the α's are constant. They can of course vary with population density and also with the abundance of resources. To demonstrate that the result is robust, that it does not depend on these simplifications, we also give an alternative proof in which the amounts of resource may change as a result of being utilized. This alters the degree of competition, but not the final conclusion.

For the purposes of this theorem, a community such as Harrison's birds which divides the habitat by food type and height in the canopy would have as many resources as there are combinations of layer and food. But immediately a difficulty arises: how similar can resources be and still count as distinct resources? How can a continuum of resource (temperature range of habitat, height, food particles size) be divided into discrete resources?

First, we note that the equilibrium depends on the superiority of each species in part of the resource space. If two resources are so similar that the same species is superior in both, only one species will persist. A more detailed analysis can procede as follows (after MacArthur and Levins, 1967):

On a continuum of one dimension, let the niches of an array of species be as in Figure 3.5. If there are only two species and the K's are equal, there is no limit to the similarity of two species short of complete identity. Now consider three species. The middle one competes to the extent α with the other two, while the outside species compete to the extent β. If the middle species is absent, the other two reach equilibrium at $x_1 = x_3 = K/(1 + \beta)$.

Then species 2 can increase only if

$$K - \alpha \frac{K}{1 + \beta} - \frac{\alpha K}{1 + \beta} > 0.$$

Thus it is required that

$$\alpha < (1 + \beta)/2.$$

If the outside species do not overlap at all, α must be less than $1/2$, and the sum of the α's less than 1. The competition between species 1 and 3 reduces their ability to compete with species 2, so that α can be somewhat greater. If the niche has a normal shape $exp\,(-x^2/2\sigma^2)$, two species whose preferred environments differ by M will compete to the extent $\alpha = exp\,(-M^2/2\sigma^2)$. Then the maximum α allowable between adjacent species is about 0.54 instead of 0.50 for a rectangular niche. From this we find the closest M, which is proportional to σ. Therefore, the number of species which can coexist is equal to the total range of resources divided by a multiple of the niche breadth. This is an upper limit, since inequalities in the K's and variability of the α_{ij} will reduce the number of species. For instance, a rare resource which is too similar to an abundant resource cannot support a different species.

As the dimensionality of the niche increases, each species has more immediate neighbors in niche space. But these neighbors also compete with each other. At the extreme when the number of species equals the number of dimensions plus one, all of the α's can be equal and any $\alpha < 1$ permits a demographic equilibrium.

If we are able to measure the α's, the community matrix approach of the previous section enables us to ask the slightly different question: how many species of roughly the kinds we have observed, i.e., drawn from the same universe of species with α taken from the same distribution, can coexist. Given the average, variance, and covariances of the α's, we can iterate equations 3.13, 3.14 to find the expected value D_n of the determinant of the community matrix. The greatest value of n for which D_n remains positive is our best prediction as to the numbers of species in the community. In Table 3.6 we have calculated D_n for different *Drosophila* communities. The results are of the right qualitative level, suggesting that these communities are more or less saturated.

So far we have limited the discussion to demographic equilibrium. Two other types of equilibrium must be considered. First, if a community is able to reach demographic equilibrium with N species present but if there are more than

N different competitors available, it is possible for there to be several alternative stable communities existing in different patches of the same environment. Then each community will be subject to occasional invasions, and we have to ask under what circumstances will a stable community be resistant to invasion? In the second model of the appendix, it is seen that if a pair of resources supporting two species is similar enough, species intermediate between the two (less specialized with respect to these resources) can outcompete the pair of them. "Similar enough" in this context means that, for a normal shaped niche the resources differ by less than 2σ on the environmental axis. Thus, although species can coexist on resources 2σ apart they are replaceable, and invasion phenomena will result in communities with species 2–5σ apart.

Finally we note briefly that there is also an evolutionary equilibrium brought about by the coevolution of species.

Brown and Wilson (1956) introduced the concept of character displacement to indicate an increase in the difference between two species in their region of contact as a response to selection for reduced competition. Where species compete for foods of different sizes, the size of a bird's bill or legs, or of a mammal's linear dimensions, may be an appropriate measure of displacement. This approach has been taken by Schoener (1965) and others.

But character displacement is only a special case of coevolution. A somewhat more general treatment can be given using a fitness set argument. For a given species, let the two axes represent the coefficients of competition α_1 and α_2 with two competitors on opposite sides along a single niche dimension. We will assume that the shape of the niche is fixed but that its peak can occur at any point. Thus each point in the fitness set corresponds to the center of a niche. Fitness is maximized for the components in question when total competition, $\alpha_1 X_1 + \alpha_2 X_2$ is minimized. Here X_1 and X_2 are the relative abundances of competitors 1 and 2.

In Figure 3.6 we show the optima. In 3.6a, the fitness set is concave. This means that species 1 and 2 are further apart than 2σ. Then the optimum strategy for an incoming

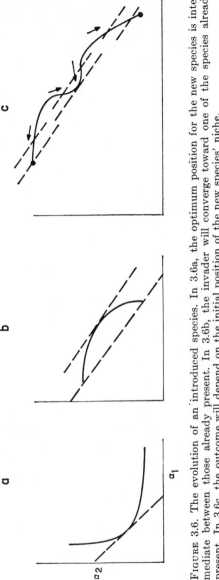

FIGURE 3.6. The evolution of an introduced species. In 3.6a, the optimum position for the new species is intermediate between those already present. In 3.6b, the invader will converge toward one of the species already present. In 3.6c, the outcome will depend on the initial position of the new species' niche.

species would be to occupy an intermediate position. If species 1 and 2 are equally abundant, it would be the midpoint. In 3.6b, species 1 and 2 are closer than 2σ, so that the fitness set is convex. Here fitness is maximized by approaching and displacing either species 1 or 2, depending on where the species starts from.

However, if the niche is multidimensional, convergence on a single axis or on several axes need not result in displacement. We could obtain clusters of very similar species in apparent violation of Gause's principle and our intuitions about natural selection.

If the niche curves are of normal shape, the fitness set will not be pure concave. Rather it will look like that in Figure 3.6c. Depending on the initial condition of the incoming species, it may approach one of the previously existing species or occupy an intermediate range.

Thus the coevolution will tend to pack species in their niche space. If the niche space is one-dimensional, the limiting packing will be to a closeness of about 2σ. For a higher dimension, with competition spread more evenly, packing may be somewhat closer. But all of our analyses agree that the number of species will depend on the range of environment divided by niche breadth.

The analysis of adaptive strategies in Chapter 2 showed that a species loses fitness in a heterogeneous environment. It would be better off specialized, but the uncertainty of the environment forces niche expansion. Thus we can assert roughly that the number of species which can be maintained in a community is proportional to the uncertainty of the environment. This may account for increase in numbers of species in areas of stable environment and high productivity, such as Hawaii and the tropical forests.

The approach outlined in this chapter leads to a number of qualitative predictions which are testable in the field. For instance, the number of species should be greater in regions of environment stability, smaller in groups with good homeostasis, smaller in new-equilibrium faunae than in old ones. It would also be used for more quantitative predictions relating the number of species to the average and variance of α's and even to get relative abundances from niche breadth and α's.

Several types of laboratory experiments suggest themselves. In *Drosophila*, a heterogeneous environment can be provided in population cages with food cups of different kinds. We could make one kind of cup easily identifiable (say by a textural or olfactory signal), and unfavorable in varying degrees (say with DDT, or simply by removing a given proportion of the pupae that emerge). Flies are collected as they hatch, released into the cage for a short time to lay eggs, and the cage is then cleaned by a vacuum cleaner or given a heat shock that removes adults but won't kill eggs. The time available is the equivalent of productivity, and the difference between the viability in the two environments is the other essential parameter. This system should result in the evolution of optimal niche breadth and specialization.

Or a community could be simulated using several types of habitat and species. The α's could be calculated first, and the equilibrium populations predicted. But these will also evolve in the course of the experiment.

Beyond question of specific prediction, this approach emphasizes the sufficient parameters niche breadth, dimension, overlap, and community diversity as objects of study. We would want to know their statistical distribution in nature, and differences between types of animals or habitats in their means and variances.

A number of theoretical questions have scarcely been investigated. For example, we don't have any useful theory as yet relating the community matrix to the relative abundance of constituent species. Another series of problems relates to ecological engineering: given a matrix, how can a minimum alteration be applied to get maximum effect, either toward eliminating a species or toward stabilizing the community.

The equations that led to the community matrix may be adequate for the equilibrium condition, but the logistic is very poor as a description of the dynamics of a community. It may be, as Kerner (1957) and Leigh (1965) believe, that the unrealistic aspects of the Voltera equations have opposite effects, and that the behavior of the whole system may not be too far from that described by their statistical mechanical approach. But that is still open to question.

APPENDICES TO CHAPTER THREE

Appendix I. The Expected Value of the Determinant of the Community Matrix

Consider the $n \times n$ matrix whose elements are α_{ij}, the coefficients of competition of species j on species i:

$$M = \begin{Bmatrix} 1 & \alpha_{12} & \alpha_{13} & \cdots & \alpha_{1n} \\ \alpha_{21} & 1 & \alpha_{23} & \cdots & \\ \cdot & & & & \\ \cdot & & & & \\ \cdot & & & & \\ \alpha_{n1} & & & & 1 \end{Bmatrix}$$

The determinant of this matrix can be expanded by minors of the first row and first column to give

$$D_n = D_{n-1} + \sum_{i,j} (-1)^{i+j+1} \alpha_{i1} \alpha_{1j} A_{11ij}$$

where A_{11ij} is the $(n-2) \times (n-2)$ determinant formed by deleting rows 1 and i and columns 1 and j of the original determinant. Now define T_n as the same determinant with the first column replaced by 1's, and expand in the same way:

$$T_n = D_{n-1} + \sum_{i,j} (-1)^{i+j+1} \alpha_{1j} A_{11ij}.$$

We now assume that all the α_{ij} are statistically independent except that α_{ij} may be correlated with α_{ji}. With this assumption, the expected values of D_n and T_n are

$$E(D_n) = E(D_{n-1}) + \bar{\alpha}^2 (n-1)^2 E\{(-1)^{i+j+1} A_{11ij}\}$$
$$- (n-1) \operatorname{cov}(\alpha_{ij}, \alpha_{ji}) E(D_{n-2})$$
$$E(T_n) = E(D_{n-1}) + \bar{\alpha}(n-1)^2 E\{(-1)^{i+j+1} A_{11ij}\}.$$

Now remove a species from the community. Since the ordering of species is arbitrary, we can remove the first species. T_{n-1} is formed by replacing the second column by 1's:

$$T_{n-1} = \begin{Bmatrix} 1 & \alpha_{23} & \alpha_{24} & \cdot\cdot\cdot & \alpha_{2n} \\ 1 & 1 & \alpha_{34} & \cdot\cdot\cdot & \\ \cdot & & & & \\ \cdot & & & & \\ \cdot & & & & \\ 1 & & & & 1 \end{Bmatrix}$$

This can be expanded by the minors of elements of the column of 1's:

$$T_{n-1} = \sum_j (-1)^{i+j} A_{112j}.$$

This has the expected value,

$$E(T_{n-1}) = E\left\{ \sum_{j=2}^n (-1)A_{112j}\right\}$$

But since the ordering of species is arbitrary,

$$E(A_{112j}) = E(A_{11ij})$$

and

$$E(T_{n-1}) = -(n-1)E\{(-1)^{j+i}A_{11ij}\}.$$

Substituting this value in the equations for D_n and T_n, and dropping the E symbol, we have the final result

$$D_n = D_{n-1} - (n-1)\bar{\alpha}^2 T_{n-1} - (n-1) \operatorname{cov}(\alpha_{ij},\alpha_{ji})D_{n-2},$$
$$T_n = D_{n-1} - (n-1)\alpha T_{n-1}.$$

When there is no covariance, this reduces to

$$D_n = D_{n-1} - (n-1)\alpha^2 T_{n-1}$$

for which we have the explicit solution

$$D_n = (1-\alpha)^{n-1}\{1 + (n-1)\alpha\},$$
$$T_n = (1-\alpha)^{n-1}$$

Appendix II. Proof That the Number of Species Cannot Exceed the Number of Resources

Proof 1: Consider the set of simultaneous equations

$$x_i' = r_i x_i \left(K_i - \sum_j \alpha_{ij} x_j \right) \Big/ K_i$$

If there is an equilibrium community other than all $x_i = 0$, the determinant of the matrix A whose elements are α_{ij} must be different from zero, which requires that the matrix be of rank n for n species. But by the definition of α_{ij},

$$A = BPP^T$$

where:

$$B = \left\{ \begin{matrix} \Sigma p_{1h}{}^2 & 0 & 0 & 0 \\ 0 & \Sigma p_{2n}{}^2 & & 0 \\ 0 & & & \Sigma p_{nh}{}^2 \end{matrix} \right\}$$

$$P = \left\{ \begin{matrix} p_{11} & p_{12} & \cdots & p_{1k} \\ \cdot & p_{22} & \cdots & \cdot \\ \cdot & & & \cdot \\ \cdot & & & \cdot \\ p_{n1} & & & p_{nk} \end{matrix} \right\}$$

and P^T is the transpose of P, with elements $p_{hj}{}^T = p_{jh}$. Here p_{ih} is the proportion of resource h (or habitat h) among the resources used by species i. Thus B is of rank $n \times n$, P is an $n \times k$ matrix when there are n species and k resources, and P^T is $k \times n$. Hence if $k < n$, A is of rank k and there can be no stable community.

Proof 2:

Let there be k resources R_h, $h = 1 \ldots k$, and n species utilizing these resources. Suppose that species i can increase as long as $\sum_h C_{ih} R_h > T_i$ where the C_{ih} are inverse measures of the effectiveness of utilization of resource h by species i and T_i is some threshold. On a graph whose coordinates are the resource levels R_h, each species can increase as long as the point $(R_1, R_2, \ldots, R_k)$ representing the combination

of resources present lies above the plane $\sum_h C_{ih}R_h = T_i$.
Where the resources are represented by a point on the plane, the species is in equilibrium, and where m planes intersect we have the set of possible resource combinations that can just support m species. Finally the intersection of k such planes defines a single point at which k species can be in equilibrium with k resources. It is infinitely unlikely that $n > k$ planes intersect at a point in k space. The $(k + 1)$st species will intersect with any $k - 1$ others at a point either closer to the origin or further away than the intersection of the first k. This defines an alternative community which will either replace or by replaced by the first one.

CHAPTER FOUR

The Species in Space

Population samples of the same species taken from different localities will, in general, be different. These differences have been described variously in terms of clines, of mean phenotypes, of clines in frequencies of genes or chromosomes, and of mosaics, races, and ecotypes. These patterns may be the results of the direct action of the environment on phenotype, the consequences of population density and age distribution, and genetic differences resulting from random drift, selection, and migration. We are primarily concerned with the following questions:

1. What is the relation between phenotype plasticity and genetic differentiation in adaptation to spatial heterogeneity?
2. How do we account for discontinuities in the spatial pattern? Is the genotypic composition a continuous, one-to-one mapping of the environment?
3. What is the role of environmental grain?
4. Do different kinds of traits have different spatial patterns?
5. What is the nature of a species' boundary?

Since the classic experiments of Clausen, Keck, and Heisey (1940) numerous studies have been carried out in which populations are compared in their own natural environments and also under controlled laboratory conditions or in reciprocal transplantations. An interesting recent study on these lines is that of Gregor and Watson (1961). They compared three species (*Plantago maritima*, *Agrostis tenuis*, and *Festuca ovina*) in many meadow habitats, classified into three main types. Populations from different vegetation types differed under uniform conditions in the same direction as they differed in their natural habitats, but less so. Thus, for *Plantago* the variance of leaf length among populations of different wild habitats was 157, but in the garden only 28. For *Agrostis* the corresponding figures were 138 and 7. Thus, in the first case 18% of the variance among populations was genetic, in the second only 5% was genetic. Clearly, these species adapt to the differences in habitats primarily by developmental flexibility.

Gregor and Watson also examined differences among populations from different microhabitats, that is, from sites within the same vegetation types. Here they found no correlation between genotypic and phenotypic variation. Nevertheless, there was a positive correlation between phenotypes of different species over sites, indicating that the phenotypic differences were adaptive. The smaller genetic variance in *Agrostis* may be a consequence of its perennial habit, which slows down the response to selection. In general, we would expect that long-generation species in variable habitats must adapt primarily by developmental flexibility. The microhabitat differences among sites may be of shorter duration than among major vegetation types, and not allow time for either species to differentiate genetically.

If this interpretation is correct, it could be verified by a comparison of annuals, perennial herbs, and non-climax shrubs and trees in successional formations. Genetic variation should be least in the latter.

Many studies of patterns of variation in morphology lead to the same conclusion: Populations differ under uniform laboratory conditions in the same direction as they do in nature. See, for example, Tantawy and Mallah (1961) for *Drosophila simulans* in the Middle East. Since the genetic variation is positively correlated with the direct environmental effects, we designate this pattern "co-gradient variation."

However, it is by no means the universal pattern. On a 15-mile transect from sea level at Guanica, Puerto Rico, to a 3,000-foot altitude at Indiera Baja, we found very little difference in the wing length or body weight of wild *Drosophila melanogaster*. The coastal population, living at higher temperatures, were only slightly smaller. But at uniform laboratory temperatures, the coastal flies were larger. This pattern, in which the direct effect of the environment acts in the opposite direction from genetic differences, is designated "contra-gradient variation." It seems to be rare for the usual morphological differences, but it is apparently the rule for geographic variation in metabolic rate in invertebrates. Vernberg (1959) cited a number of such cases in marine organisms. Generally the metabolic rate of the tropical form is higher than that of the temperate form if each is compared at its own temperature, but lower when both are tested at the same temperature. This pattern is further complicated because temperate and tropical forms are not equally sensitive to temperature variation, so that their oxygen consumption/temperature curves may cross.

These patterns can be explained most readily by relaxing the assumption that there is no necessary relation between the physical form of the signal and the response. While photoperiod can be linked almost arbitrarily to any kind of flowering or dormancy response, temperature acts more directly on basic physico-chemical processes. It seems to be a universal characteristic of invertebrates that size decreases with temperature although the Q_{10} may vary. If we assume a fixed temperature/size relation but assume that the adaptive significance of size is more directly related to desiccation in *Drosophila*, the optimum size will not have any necessary relation to temperature. In Figure 4.1, we show how a fixed temperature/size relation can produce co-gradient variation when high temperature is correlated with high humidity as in Tantawy's transect, and contra-gradient variation when the hot regions are also the dry ones, as is the case in our transect.

In the previous analysis the spatial pattern would be expected to be in continuous one-to-one correspondence with the environment. However, this need not be the case. We first treat the spatial pattern by means of the fitness set strategy, and then consider some genetic aspects. Return now to the fitness set diagrams of Figures 2.3. The parameter p (which represents the proportion of habitat I in a local

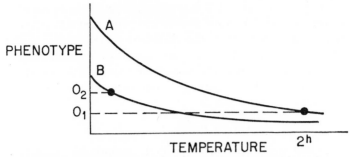

FIGURE 4.1. Cogradient and contragradient geographic variation. The solid lines show the relation of body size to temperature as a direct consequence of development. O_1 and O_2 are optimum phenotypes for the actual temperatures of the populations represented by curves A and B. The optimum is achieved by altering the intercept of the solid curve upward or downward. The optimum is shown to increase with temperature, so that populations in hotter regions are genetically larger. If the increase is moderate, they may still be smaller in nature.

mosaic or the probability of environment I in a temporally variable situation) can vary geographically. For the convex fitness set of Figure 2-3a, where the alternative environments are not too different, the optimum phenotypes are found to vary continuously with p. However, the rate at which the optimum changes with p depends on the curvature of the fitness set. We see that the steepness of the population cline will not be a simple function of the steepness of the environmental gradient.

When the alternative environments are very different compared to individual tolerance, the fitness set is concave. If the environmental pattern is fine-grained, the Adaptive Function will be linear,

$$A = pW_1 + (1 - p)W_2, \qquad (4.1)$$

where p is the proportion of habitat I. This would result in local specialization, with the phenotype which is optimal in habitat I when p is more than $1/2$; and changing abruptly to the phenotype adapted to environment II when p falls below $1/2$. The spatial pattern along a transect in p is thus one of discrete races. Of course, the boundary in nature would be smoothed over to some extent by migration. (The threshold value $p = 1/2$ is a consequence of the symmetry of the model. If optimal fitness in environment II is twice that in environment I, the threshold would be $2/3$.) The important point here is that the zone of transition need not correspond to a region of abrupt environmental change.

If the patchiness of the environment is coarse-grained, the Adaptive Function is

$$A = p \log W_1 + (1 - p) \log W_2. \qquad (4.2)$$

Now the optimal population will be polymorphic, a mixture of types specialized to the two habitat phases. But the proportion of the morphs will vary in a continuous way along a cline in p.

Real fitness sets would be expected to be convex near the ends and concave in the middle. The result of this is a pattern which is part monomorphic cline, part polymorphic mixture. The zone of polymorphism might be misinterpreted as a region of hybridization.

The above analysis, taken from Levins (1962), considers only the optimal strategy subject to constraints on the fitness of individual phenotypes. It may be that for polygenic inheritance an appropriate genetic system may arise which gives the optimum pattern. However, at least when fewer loci are involved, the genetic system imposes additional constraints on the optimum strategy. We must now consider how this modifies the theory.

The spatial pattern of genotypic variation in a species is the result of the interaction of natural selection, migration, and random drift. It is clear that drift produces erratic differences in gene frequency independent of the environment and that migration reduces the differences between populations. The effects of selection are more complex and will be treated first alone.

At first glance, it might seem that selection, acting alone, would give rise to a spatial pattern which varies in a continuous, one-to-one way with the environment. This is most conveniently tested using the fitness set analysis introduced in Chapter 2. In Figure 4.2 we show a situation for a single locus and two alleles. The heterozygote is taken to be

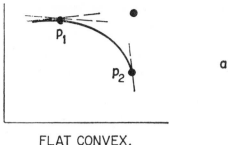

FLAT CONVEX.

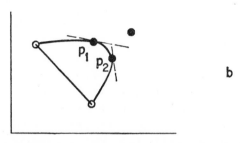

FIGURE 4.2. The steepness of a cline as a function of the convexity of the fitness set. In 4.2a, a flat fitness set results in rapid geographic variation. In 4.2b, a very convex fitness set allows small phenotypic changes to cope with wide environmental variation.

superior to each homozygote in one environment, inferior in the other, and its average fitness exceeds that of the homozygotes so that its fitness point lies up and to the right of the line joining the homozygotes. The curved line represents all possible mendelian populations.

The type of selection depends on the "grain" of the environment. A fine-grained environment occurs in small units; the individual encounters many in its lifetime; therefore, the effective environment is an average of these units. In such cases, selection will maximize $pW_1 + (1 - p)W_2$, where p is the frequency of environment I and W_1 and W_2 are fitnesses in environments I and II. Thus, the population will reach equilibrium at the gene frequency corresponding to the point of tangency of the straight line

$pW_1 + (1 - p)W_2 = K$ with the fitness set. We note the following consequences of this model:

1. If p is too close to zero or one, the population will be monomorphic (homozygous) and specialized to the more abundant habitat. (What is meant by "too close" depends on how great the average heterozygote superiority is. If the heterozygote is superior to both homozygotes in both environments there will always be polymorphism.) Since for intermediate values of p the measure of environment diversity (see Ch. 3) is greatest, we can assert that the degree of polymorphism increases with habitat diversity, a result already shown by Levene (1953) for a coarse-grained environment.

2. If the fitness set is flat, as in Figure 4.2a, there will only be a narrow range of environmental proportions (and, presumably, a narrow spatial range) over which polymorphism is stable. But in this range gene frequency passes rapidly from zero to one, so that this local population will be strikingly more heterozygous than populations in other parts of the species' range. This will usually be regarded as a zone of hybridization, when in reality it could be simply a region of maximum environmental diversity. Such diversity is most likely to occur at the borders of vegetation or soil types or climatic transition zones, or disturbed habitat, precisely where zones of contact and hybridization would also be expected.

3. It is clear that the rate of change of gene frequency with p depends on the shape of the fitness set. Where the set is flat, a small change in p would result in a large change in gene frequency, whereas where the fitness set is sharp, a large change in p may result in only small changes in gene frequency. For example, in Figure 4.2b the whole range of p values from zero to one moves the gene frequency from p_1 to p_2. Hence, we can have both steep and shallow clines along the same environmental gradient. A low level of geographic variation in a species need not mean that there is little selection or that all genotypes are more or less equal.

Among the types of polymorphism which have been

most extensively studied are those for chromosomal inversions in *Drosophila*. While some species have the same inversions over a very large area (e.g., *D. subobscura* in Europe) others are restricted to regions much smaller than the species' range. (This is more true in the *D. willistoni* group.) We suggest that the fitness set analysis may show *D. subobscura* and *D. willistoni* to differ in the way that Figure 4.2b differs from Figure 4.2a. Roughly, the geographic range of an inversion polymorphism should increase with the average heterosis.

In a coarse-grained environment, each individual spends its whole life in one of the possible environment types. Then selection maximizes $p \log W_1 + (1 - p) \log W_2$. (See Ch. 2 for more detailed discussion.) On a convex fitness set of Figure 4.2 the results are roughly the same. However, where in a fine-grained environment the slope of the Adaptive Function is

here it is

$$\frac{\partial W_1}{\partial W_2} = \frac{-(1 - p)}{p}, \qquad (4.3)$$

$$\frac{\partial W_1}{\partial W_2} = \frac{-(1 - p)}{p} \frac{W_1}{W_2}. \qquad (4.4)$$

Hence, where $(1 - p)/p$ is large, W_1/W_2 is small and vice versa. Thus $\partial W_1/\partial W_2$ is held more to an intermediate range of values. The region of polymorphism is broader, and the change of gene frequency with p is slower. Further, this effect depends on the maximum ratio W_1/W_2. If the two environments are very different, then near homozygosity W_1/W_2 will have extreme values, while for the environments more similar this effect is reduced.

Now consider the situations shown in Figure 4.3. Here the heterozygote is on the average inferior (its fitness point is below and to the left of the line joining the homozygotes). In a fine-grained environment there can now be no polymorphism. The population will be homozygous for one allele over a wide range of values of p, and when p passes some threshold will jump abruptly to the other type. This corresponds to the purely ecological concave fitness set

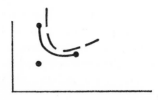

FIGURE 4.3a. Stable polymorphism with a concave fitness set.

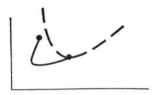

FIGURE 4.3b. Alternative fixations when the heterozygote is very inferior, resulting in race formation.

FIGURE 4.3c. A mixed case, with one polymorphic and one monomorphic optimum.

analysis of Chapter 2. It is a departure from the hypothesis of a continuous mapping of population on environment.

In a coarse-grained environment some new phenomena arise. If the curvature of the fitness set is less than that of the Adaptive Function, as in Figure 4.3a, there will be a stable polymorphism. This is the situation treated by Levene. If the heterozygote is more inferior, as in Figure 4.3b, the curvature of the fitness set is increased. Then what was at first a point of stable equilibrium becomes unstable, and there are two stable situations. In this figure both are homozygous, but there can also be one polymorphic and one monomorphic equilibrium, as in Figure 4.3c. This is the first time that we have found alternative steady states.

The population will go to one or the other depending on its past history. If it enters the region from an area of high gene frequency of A, the high A stable point will persist. If it enters from a low A region, it can produce a stable population without any A. If the species enters the region from both sides, it can produce two kinds of populations corresponding to the two stable equilibria. Their subsequent fates will depend on the extent of migration among populations. For slowly moving species such as snails it could result in a patchwork of phenotypically distinct populations which do not correspond to differences among their environments. Since the selective values of other genes may depend on the frequencies at the A-locus, an initial population difference due to historical causes may be amplified by further selection.

Clarke (1966) offers an interesting model to account for a curious pattern of variation in the snail *Cepaea nemoralis*, in which "the predominance of a few morphs, apparently regardless of habitat, characterizes areas much larger than that of a panmictic population. Between such areas the morph-frequencies may change violently over distances of 200 meters or less." He considers an environmental cline acting on a heterotic locus to produce a cline in the frequency of a gene, A. A second locus, B, shows simple dominance, but the selective values depend on the frequency of A and in turn affect those frequencies. Then along the cline in the frequency of A there will be two regions showing respectively B-frequencies of zero and one. Further, the presence of B can increase the relative fitness of A and thus steepen (or flatten) the cline at a point that does not correspond to an environmental transition. Here an initial continuous environmental gradient produces discontinuities in the distribution of genes, but unlike the previous situation, there is an ultimately one-to-one correspondence to environment (although not continuous) which is stable. The process can, of course, be repeated, building up complex genetic differences in space.

The shape of the fitness set depends on how different the alternative habitats are compared to the range of tolerance of the individual genotype. If the environments become more similar or if homeostatic capabilities increase, the homozygous fitness points approach each other. Where the heterozygote was superior (Figure 4.2), this makes the set more convex upward. Hence, 4.2a approaches 4.2b. Then the geographic extent of stable polymorphism increases, but gene frequencies will vary less over the range. This is true of both coarse-grained and fine-grained environments. Where the heterozygote was originally inferior, we do not know what would happen. For a fixed degree of inferiority the fitness set becomes more concave. But we might expect the heterozygote to become less inferior, thus having the opposite effect.

The fitness set is flattened by inbreeding. If it was originally convex, inbreeding makes it less so, reducing the belt of equilibrium polymorphism and making the population vary more from place to place.

In the case of an inferior heterozygote, inbreeding reduces the curvature and therefore can turn an unstable into a stable polymorphic equilibrium. For the purposes of this discussion, genotypic assortive mating would have the same effect. This increase in inbreeding also increases fitness. There are two consequences of interest:

1. There is a selective advantage to inbreeding or genotypic assortive mating for those traits that have a concave fitness set, and a disadvantage for those which have a convex fitness set (heterozygous superiority). This is true even if the different genotypes do not show preferences for the environments in which they do best. The combined effect of selection for and against assortive mating would seem to lead to an optimum level, which could be expressed as partial sexual isolation among sympatric strains within a population.

The optimization of F can be seen most readily in a symmetric situation where $p = 0.5$, the homozygotes have fitnesses $W_1(aa) = W_2(AA)$ and $W_2(aa) = W_1(AA)$, and the heterozygote has the fitness $\frac{1}{2}\{W_1(aa) + W_2(aa) + h\}$ in both environments. Here h may be positive or negative, and measures mean heterosis or heterozygous inferiority. Inbreeding moves the fitness set so that the point corresponding to gene frequency 0.5 is displaced up or down by the amount Fh. This increases or decreases W_1 and W_2 by $Fh/\sqrt{2}$.

For a fine-grained environment,

$$\frac{\partial \bar{W}}{\partial F} = -h/\sqrt{2}. \tag{4.5}$$

For a coarse-grained environment,

$$\frac{\bar{W}}{F} = \frac{h}{\bar{W}\sqrt{2}} = \left(\frac{-h}{\bar{W}_0 + hF}\right)\Big/ \sqrt{2} \tag{4.6}$$

where $\bar{W}_0$ is fitness at zero inbreeding. Suppose that q_1 of the loci undergo fine-grained selection with heterosis, q_2 undergo coarse-grained selection with heterosis, and $q_3 = 1 - q_1 - q_2$ undergo coarse-grained selection with heterozygous inferiority. Let all h's be the same.

$$\frac{\partial \bar{W}}{\partial F} = \frac{1}{\sqrt{2}} \left\{ -q_1 h - \frac{q_2 h}{W_0 + hF} + \frac{q_3 h}{W_0 + hF} \right\}. \tag{4.7}$$

Thus if $q_3 > q_2 + W_0 q_1$ the optimum F is

$$\hat{F} = \frac{q_3 - q_2 - W_0 q_1}{q_1 h} \tag{4.8}$$

But the level of optimum inbreeding varies geographically and may completely disappear. Thus, these sympatric strains will be part of a panmictic population in some regions and will behave almost as good species in others. They need not represent incipient species. From the argument of habitat selection, it seems that in a highly productive environment the genotypes would preferentially seek the habitats or resources in which they do best. This would lead to the accumulation of other associated adaptive differences and could lead at least locally to speciation. But where the environment is more uncertain this specialization will not be favored. It is obvious that this system can give rise to complex species with regions of heterogeneity and strong

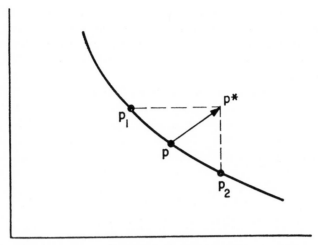

FIGURE 4.4. The consequences of habitat preference. With a mean gene frequency giving the point p on the fitness set, in environment I the frequency is p', and in environment II it is p''. This displaces the fitness point up and to the right to p^*.

differentiation, regions of panmixia, and regions of monomorphism. This could mislead taxonomists into seeing points of contact between independent species.

2. The amount of inbreeding will depend partly on the size of the habitat patches for coarse-grained environments. In the original Levene model, individuals from all patches mingled freely to mate and laid their eggs at random with respect to patch. If mating takes place soon after emergence, it is possible to mate preferentially with individuals from the same patch and yet deposit eggs at random. Then the zygotic frequencies are equal in all habitats and the only effect is to flatten the fitness set. Since inbreeding increases with patch size, we conclude that there may be a minimum patch size for polymorphism—rare patches which are too small will not affect population composition. If individuals tend to stay in the same patch in which they were born, gene frequencies will not be the same in all habitats. In Figure 4.4 we show that this flattens the fitness set further. It approaches a situation better handled as distinct populations with migration.

Circumstances which give rise to alternative stable equilibria require further discussion. We have already shown that in a coarse-grained environment there can be more than one stable equilibrium even at a single locus. Since each equilibrium point corresponds to populations whose fitnesses are greater than those of slightly different populations, Sewell Wright has referred to them as adaptive peaks on a landscape whose height is fitness and whose other dimensions are gene frequencies. The occurrence of multiple adaptive peaks makes it possible for different species to coexist without converging. Their occurrence within a species means that even when populations are in equilibrium determined completely by selection there is not a one-to-one correspondence of population composition and environment. Such multiple peaks can arise in the following conditions:

1. The same phenotype can be achieved by alternative sets of genes.

   (a) Isoalleles may produce enzymes which function equally well despite subtle differences detectable in the laboratory.
   (b) In a multi-locus system, polygenes may all function additively to produce some phenotypic trait. If fitness, itself, were additive, there would be only a single equilibrium—the fixation of the more fit allele at each locus. But suppose that there is an optimum phenotype S, and that each locus contributes $\pm S/5$ to the phenotype. Then if we start with 9 loci, an optimum can be reached by fixing the $+S/5$ allele at any 7 loci and the $-S/5$ allele at the other two. This may be done in $7 \times 6/2 = 21$ different ways corresponding to 21 equal adaptive peaks. Of course, we needn't assume fixation at all loci.

2. Gene frequencies at different loci need not be in linkage equilibrium. Lewontin (1964) showed for a two-locus model that there may be two stable equilibria, one with

79

linkage equilibrium as well, the other with an excess of repulsion over coupling gametes.[1] The alternative peaks are not generally of equal fitness.

3. Adaptive problems may be solved with alternative phenotypes. Thus, the researches on color and band pattern in snails of the genus *Cepaea* (Clarke) indicate that their adaptive significance is related to camouflage. In *Cepaea nemoralis*, a decrease in the light-yellow ground color of the shells is associated with darker habitats. In the closely related *C. hortensis*, the same effect is achieved while retaining the yellow background but fusing adjacent dark bands; there are many other groups in which color patterns vary widely and even separate species, although there is no obvious superiority of one over the other. King (1965) has shown that DDT resistance may be achieved independently in different *Drosophila* populations by different means, with different genes and different correlated changes.

Miss Angelica Muniz, in our group, has been studying geographic variation in the spider *Gasteracantha tetracantha* in the Puerto Rican region. The width of the spider's body shows a regular pattern with a maximum in the Puerto Rican highlands and minima on the islands of Desecheo, Culebrita, and Anegada. There is no indication of any long-range clines or distance effects, and determination would seem to be purely ecological.

These spiders also show striking polymorphism in abdominal color patterns, which may be all black, black and yellow, or black and white. There were 19 patterns recognized for females. Some were found throughout the region, while others were endemic to Desecheo Island, to Puerto Rico, or to other groups of islands. Thus, it is possible to compute coefficients of similarity between islands using Preston's (1962) coefficient $(1 - Z)$ which was designed for comparing ensembles of species. Roughly, these coefficients corresponded to distance. Thus, the St. Thomas area had similarities of 0.57 and 0.53 with its eastern and western neighbors, 0.47 with the more remote Virgin Gorda, and 0.36 with Anegada at the end of the chain. This pattern

---

[1] Repulsion gametes have the form Ab or $a\beta$.

TABLE 4.1. Variation in color polymorphism in *Gasterocantha tetracantha*

Region	Per cent		
	White	Yellow	Black
Desecheo	5	90	5
Puerto Rico	10	67	21
Small islands, between P. R. and St. Thomas	52	43	4
Hans Lollick and Little Hans Lollick	22	78	0
Tortola group	58	40	2
Salt, Ginger, and Norman Islands	57	38	0
Virgin Gorda group	59	39	2
Anegada	71	29	0

would, therefore, seem to be a classical isolation by distance situation, with historical rather than ecological determination of distribution.

However, let us assume that color pattern is related to either heat absorption or camouflage. In either case we would expect an abundance of dark forms in the rain forests and lighter morphs in open environments. Table 4.1 shows the distribution of the abdominal patterns grouped by color. If those white or yellow types whose body is more than half black are included as black, the situation is not basically altered. We see that black reaches a maximum in Puerto Rico and is low on the smaller islands. The light color is mostly yellow on Desecheo and is gradually replaced by white as one goes to the east. Desecheo lies outside the Puerto Rico shelf although only some 12 miles west of Puerto Rico. The other islands were all connected to each other during the glacial maxima. Thus, we find that historical events and distance may determine which morphs are present, but the proportion of light and dark types is ecologically determined.

4. The adaptive significance of a trait may depend only on its being common rather than on its special qualities in relation to environment. This is true of the recognition signs necessary to mating, of courtship patterns, and of bird calls and songs. Here the first genes to be abundant in a population have an advantage over any variants.

81

In general, an "adaptive landscape" depends on the environment, so that it varies from place to place. Populations in different localities may be occupying corresponding peaks which have been displaced somewhat, or they can be on totally different peaks. In the former case, differences among populations correspond in a one-to-one way to differences among their environments: selection practiced on any one of the populations can bring it to approximate any other in relatively few generations; each one would respond to new selection in the same way, and crosses between populations are likely to produce intermediates in the F_1 with perhaps a small increase in F_2 variance around the same mean. We define a region whose populations differ in this way as a genetically coherent region, and the population as a genetically coherent series.

In contrast, populations which do not belong to the same genetically coherent region do not differ in ways showing one-to-one correspondence to environmental differences. Ordinary phenotypic selection will not readily transform one into the other, and they are likely to respond differently to new selection pressures. Crosses between such populations may not be intermediate; in any case, since fitness depends on the joint effects of combinations of genes, we would expect a breakdown in the F_2 of the sort described by Wallace and Vetukhiv (1955).

We suggest that the genetically coherent series may be the major subdivision of the biological species, since it is based directly on the processes of differentiation and adaptation instead of only their morphological consequences. However, the extent to which the different criteria of genetic coherence delimit the same series is an open one for experiment and observation.

The problem arises, under what conditions can a population go from one genetically coherent series to another? Consider first a simple model with one locus and two alleles. Selection will be taken to favor the more prevalent allele, so that the change in gene frequency due to selection alone would be

$$\mathrm{d}x/\mathrm{d}t = -Sx(1-x)(1-2x). \qquad (4.9)$$

The result would be fixation at $x = 0$ or $x = 1$, depending on whether x begins below or above 0.5. Now we introduce random drift. The change in gene frequency can be treated as a diffusion process by techniques made familiar mostly by Kimura (1964). The gene frequency will have some probability distribution that changes with time. Eventually it will approach a steady state given by

$$\phi(x) = \frac{1}{V} \exp 2 \int \frac{M}{V} \, dx \qquad (4.10)$$

where M is the mean change in x starting at x, and V is its variance. Here

$$M = -Sx(1 - x)(1 - 2x)$$

and

$$V = \frac{x(1 - x)}{2N}$$

for populations of size N. This gives

$$\phi(x) = \frac{C}{x(1 - x)} e^{-4SNx(1-x)} \qquad (4.11)$$

where C is a constant such that $\int \phi(x) \, dx = 1$. But $\phi(x)$ goes toward infinity at $x = 0$ or 1. This means that the final result is fixation, with $\phi(x)$ giving the steady state distribution of unfixed classes. The factor $C^{-4SNx(1-x)}$ is equal to 1 at $x = 0$, 1 and goes to a minimum of C^{-SN} at $x = 0.5$. The smaller this minimu, the more the gene frequencies remain near their fixation values and the more strongly the genetic coherence is maintained. However, we would prefer a model with a true steady state distribution, where the possibility of disrupting the genetic coherence of the species is permanent. Therefore we allow enough mutation to prevent fixation. M now becomes

$$M(x) = -Sx(1 - x)(1 - 2x) + U(1 - 2x). \qquad (4.12)$$

Then the distribution becomes

$$\phi(x) = C[x(1 - x)]^{4UN-1} e^{-4SNx(1-x)}. \qquad (4.13)$$

If $4UN > 1$, so that there is an average of at least half a mutation per generation, the distribution is a true steady state. It will have peaks close to 0 and 1 and a minimum at $x = 0.5$. The value of $\phi(x)$ at that minimum indicates the rate of crossing the boundary from the region of one adaptive peak ($x = 0$) to the other ($x = 1$). This minimum depends on e^{-SN}, so that it is reduced by strong selection and large population size. Suppose, for example, that there is moderate selection ($S = 0.1$) and population size ($N = 100$). Suppose further that $4UN - 1$ is very small (arbitrarily close to zero). The ratio of the height of the peaks to the minimum is e^{10}, greater than 10^3. Thus, a transition would occur less often than once in a thousand generations. If S were 0.5, the ratio would be e^{50} and transition would already be negligible. Thus, populations with this structure would be quite stable around their peaks.

We must now introduce migration. For simplicity assume a large population with $x = 0$ and a smaller one exchanging genes with it at rate M. Then the average change in gene frequency in the small population is

$$M = -Sx(1 - x)(1 - 2x) + U(1 - 2x) - Mx, \quad (4.14)$$

and the new steady state distribution is

$$\phi(x) = C[x(1 - x)]^{4UN-1}e^{-4SNx(1-x)}[1 - x]^{4MN}. \quad (4.15)$$

The additional factor $(1 - x)^{4MN}$ reduces the possibility of the small population lying in the alternative genetically coherent series. Some examples of $\phi(x)$ are shown in Figure 4.5. It seems as if a species which is expanding slowly over a continuous area would readily preserve its genetic coherence. But if the habitable part is small and patchy

$$\phi(x) = \quad$$

FIGURE 4.5. The distribution of gene frequency of a population near a genetically coherent region of gene frequency zero.

(so that migration is small and drift large) local populations may arise which belong to the opposite genetically coherent series and persist. This may be a transient phenomenon if populations are very small and migration not too low. But if the incoherence persists long enough it can determine the fixation of other loci in opposite directions and compound the differentiation. Further, since the random events may occur independently at different loci the deviant populations will not be readily classified into groups.

A species boundary may be one of several types, each with its own genetic and ecological consequences:

1. A sharp, absolute barrier such as the sea for terrestrial species. Here marginal populations need not differ from central populations at all.

2. For organisms whose fitness set is convex, environmental change will result in phenotypic variation in a bicontinuous, one-to-one pattern. But for those with a concave fitness set and fine-grained selection the marginal phenotypes will be the same as interior ones, but they will be less successful and the species will be rarer. It will disappear when the habitat to which it is specialized becomes too rare. This kind of boundary occurs when the gap from one environment to the next exceeds the tolerance of the species and produces a concave fitness set.

Intermediate phenotypes here would be inferior to either specialized type, so that a concave gap cannot be crossed by ordinary selection of polygenes.

3. The productivity of the environment decreases progressively toward the margin because favorable patches become rarer. As argued in Chapter 3, this would result in a marginal niche expansion. As the distances between local populations increase, the probability of migration decreases, and the genetic coherence of the species is more readily broken.

4. A general deterioration of conditions makes small environmental differences at the margin more important than at the center. This could turn convex fitness sets into concave fitness sets and promote specialization.

Thus, different processes which we expect at margins of

distribution lead to opposite predictions. The theorems are not robust and an attempt to predict the qualities of marginal populations would merely illustrate the folly of seeking to resolve by the light of reason alone what will only prove amenable to knowledge. Once we know which process is dominating at the periphery, we can, of course, be more specific in predicting.

As animals increase in size, or mobility, some of the coarse-grained environmental differences become fine-grained. They cease to provide the possibility of alternative equilibria, any advantage of inbreeding which they may have contributed disappears, and polymorphism depends on heterosis. All this tends to reduce genetic variability locally.

Other, fine-grained, heterogeneities of the environment disappear completely. If a small animal wanders among patches, several meters across, which differ in temperature, these temperatures form a fine-grained pattern. But a larger animal may not reach temperature equilibrium in these different patches. The averaging of the different temperatures occurs externally.

But other coarse-grained differences arise. The home range may cover a quarter acre. One quarter acre of forest is less different from another quarter acre in the same forest than are randomly chosen square centimeters. Thus, there is more geographic heterogeneity compared to microhabitat variations.

Finally, as we pass to organisms with more elaborate homeostatic systems the fitness sets become more convex. We have already seen that the more convexly curved the fitness set, the less genetic change is produced by a given environmental difference. Therefore, the larger or more homeostatic groups will show less local variation compared to geographic variation, more clinal genetically coherent variation than formation of distinct genetically coherent series, and more phenotypic variation induced by the environment as against genetic differences.

We have indicated how various aspects of the environmental pattern and genetic system affect the spatial structure of a species. But the environmental pattern is not that of the gross meteorological environment. We have emphasized that the environmental pattern depends on the biology of the species, that environmental heterogeneity is measured relative to the tolerance of diverse conditions by the individual genotype, that it is modified by habitat selection, that "grain" depends on the size of the individual, etc. Now we will run through several groups of organisms to see how their over-all patterns would be expected to differ.

It is well known that the microspatial heterogeneity of the environment increases as the size of the organism decreases. In open vegetation, soil temperatures in exposed patches may be as much as 20°F greater than the temperature under the thick leaves of a plant. Opposite sides of the same tree differ enough to be quite different environments for mosses and lichens. Droplets of animal dung differ chemically and provide habitats for different micro-organisms. Martinez Viera (1964) reports differences in the soil fungi and bacteria in the rhizosphere around the roots of corn plants from those in the rest of the soil. Some of these differences of environment, such as temperature, also vary on a macrogeographic scale. The temperature range within a few yards of forest floor may be as great as the average difference caused by several thousand feet of elevation or hundreds of miles of horizontal distance. Therefore, for very small organisms much of the environmental diversity which occurs over large areas can be found within a small neighborhood. We have seen that for *Drosophila* the microhabitat component of diversity in species composition is at least as great as that over the whole of Puerto Rico, and that the diversity at a single site in Austin, Texas, is already about three-fourths of the total diversity over the southwestern United States. These interspecific considerations most likely apply within the species as well. We expect the coefficients of variation for traits within populations to be as great, or greater, than those among populations. Many small organisms exploit these microhabitat differences in a coarse-grained way, living their whole lives in a single host plant, rotten log, or under a single rock. Differences among

87

fruits on the forest floor are coarse-grained differences for *Drosophila* larvae, which usually develop within a single fruit, but are fine-grained for adults. Since it is in coarse-grained selection that inbreeding may be advantageous, we expect that for the small, relatively immobile invertebrates there will be a maximum of formation of partly isolated "quasiraces" living more or less sympatrically over wide geographic ranges.

More mobile small invertebrates experience more of the environmental variation as fine-grained. Here polymorphism requires at least average heterosis, concavity of the fitness set leads to fixation of genes and specialization, and alternative adaptive peaks are less common. Where different parts of the life cycle differ in mobility and hence in environmental grain, we expect polymorphism to be controlled more by the immobile stages. Thus, in *Drosophila*, we expect larval viabilities to be responsible for more polymorphism than are adult phenotypes.

If the approach developed in this chapter is valid, it can be applied as follows:

1. The notion of genetic coherence, deliberately left ambiguous at this stage of the investigation, should be explored systematically and studied as a sufficient parameter.

2. The varying steepness of clines and the occurrence of discontinuities can be studied by fitness set analysis.

3. The taxonomic significance of regions of increased variability (usually regarded as zones of secondary contact) and of partial reproductive isolation (often interpreted as incipient speciation) requires re-examination.

4. Predictions can be made as to the different geographic patterns of groups differing greatly either in their environmental patterns (grain) or flexibility.

CHAPTER FIVE

The Genetic System

Population genetics began with the study of changes in gene
frequency at a single locus in a random mating population
under the joint action of selection, mutation, random drift,
and migration. The major results were usually the establish-
ment of conditions for equilibrium. Subsequent work has
made the models more complex by allowing many loci,
arbitrary epistatic interactions among them, effects of
linkage, diverse systems of mating, fluctuations in selection,
dependence of selective values on gene frequencies, etc.

The growing availability of high-speed computers has
made it possible to simulate situations that cannot be
handled analytically, so that great masses of data are
accumulating. But it is not obvious what models to consider,
how to specify the gene interactions, or even what results to
look for. In some cases, the model is determined by a
specific biological problem—a species to be analyzed. But
for general theoretical work, the problem is, what are the
sufficient parameters of complex genetic systems which are
relevant to ecological evolutionary studies?

In the purely ecological analysis of adaptive strategies
we looked at optimum phenotypes or optimal polymorphic
populations. Where the optimum strategy is an individual
phenotypic trait such as size, degree of habitat selectivity,
or photoperiod threshold for the initiation of diapause it
can be selected for like any other phenotypic trait. One of
the restrictions is that with a continuum of possible phenotypic values but discrete genes there may not be any single
genotype which corresponds to the optimum, or if it does
exist it may be heterozygous and therefore segregate non-
optimal types. This difficulty becomes less important as

a   b

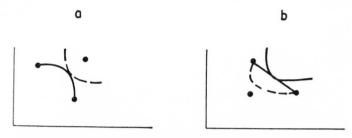

FIGURE 5.1. The genetic contraints on optimality. (a) The phenotypic optimum is a heterozygote, so that polymorphism is imposed by segregation. (b) The optimum would be a mixture of homozygotes, but this creates heterozygotes.

the number of loci increases. But it must be borne in mind that not all stable polymorphisms corresponds to an optimum strategy. Conversely, a mixed strategy may be optimal, but if it involves a mixture of homozygotes it cannot be achieved. Figure 5.1 illustrates both of these situations.

A second type of constraint involves the adaptive value of genetic change. In an engineer's adaptive system information about the environment is collected and analyzed. The record of previous experience can be designated the state of the system. On the basis of this information an optimum strategy is computed which becomes the "output" of the system. But in population genetics the output and state of the system are confounded. The information about past environments is stored as gene frequencies and gametic frequencies, but the strategy itself is also an array of genotypes. This means that a mixed strategy which requires genetic heterogeneity imposes the possibility of changing frequencies in response to selection even if the optimum strategy should always be the same (when there is no correlation between the environments of successive generations). Conversely, the capacity to respond to selection requires genetic variance even if the optimum strategy is always monomorphic.

The array of gene frequencies constitutes the "memory" of the system. Different genetic systems may have longer or shorter memories. For instance, consider additive selection as follows:

Genotype	Frequency	Fitness
AA	x^2	1
AA^1	$2x(1 - x)$	$1 - s$
A^{1}A^1	$(1 - x)^2$	$1 - 2s$

$$\bar{W} = 1 - 2s(1 - x) \qquad (5.1)$$

so that the rate of charge of gene frequency is

$$\frac{dx}{dt} = s(t)x(1 - x) \qquad (5.2)$$

where $s(t)$ is a random variable.

Here we can integrate to directly obtain

$$\log\left(\frac{x}{1 - x}\right) - \log\left(\frac{x_0}{1 - x_0}\right) = \int s(t)\, dt. \qquad (5.3)$$

Thus the present gene frequency depends on the environments of the past, all with equal weight. This is the longest memory system we can have. But by virtue of its long memory it cannot track the recent environment.

The simplest way to reduce memory is by mutation. Let there be mutation in both directions with equal frequency u. Then equation 5.2 becomes

$$\frac{dx}{dt} = sx(1 - x) + u(1 - 2x). \qquad (5.4)$$

Let the dependence of $x(t)$ on $s(t - \gamma)$, some environment of the past, be designated $Z(t,\gamma) = E[x(t)S(t - \gamma)]$, where the symbol E means expected value. Then

$$Z(t + h,\gamma) = E\{x(t)S(t - \gamma) + h[S(t - \gamma)S(t)x(t)[1 - x(t)] \\ + US(t)[1 - 2x(t)]]\}. \qquad (5.5)$$

For situations in which $E(s) = 0$ and there is no auto-correlation of the $s(t)$ with $s(t - \gamma)$ we get

$$Z(t + h,\gamma) = Z(t,\gamma) - 2uhZ(t,\gamma). \qquad (5.6)$$

This leads to the differential equation

$$\frac{dz}{d\gamma} = -2uZ \qquad (5.7)$$

which means that the dependence of $x(t)$ on the environment at $t - \gamma$ falls exponentially. The factor $2u$ is the rate at which memory is destroyed. Different genetic systems have different, usually non-linear, memory functions. Even the case we treated as a differential equation loses memory if it is replaced by a generation model.

The paradox which now emerges is that only a system with short memory can follow the environment. But the optimum parameters of the tracking system depend on the mean, variance, and autocorrelation of the environment. These can only be estimated accurately by a system with a long enough memory so that the law of large numbers operates. Since the statistics of the environment are also subject to change, the calibrating system cannot have infinite memory. There is some optimum level of memory for it, which can only be established by systems with longer memory, etc. Lewontin (1966) has used the term "capricious" to describe a random process with limited memory.

Thus the memory of different genetic systems, especially those involving multiple loci, becomes an important sufficient parameter.

The time/response pattern of multiple genetic systems has other aspects. Unpublished computer simulation studies by Lewontin and his collaborators showed that the response to directional selection may be delayed at the phenotypic level while internal changes occur in linkage relations. The delay seems to depend on tightness of linkage, and kinds of dominance, and epistatic relations. A system which waits five generations before responding can be resistant to ephemeral fluctuations of the environment and yet respond well to long-term changes. It would serve as a "filter," discarding the short-period changes of the environment as "noise."

Another aspect of optimization which is poorly understood is second-order selection. We define second-order selection as selection for a trait which does not appear in the equation for $\bar{W}$ directly, such as mutation rate. Levins (1967) showed that at a single locus with a heterotic lethal, when the fitness of the viable homozygote fluctuates the

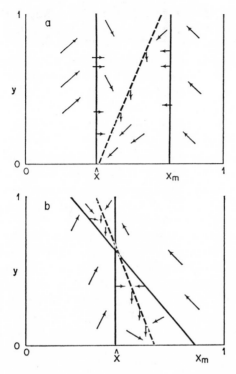

FIGURE 5.2. Selection for mutation rate. The abscissa is the frequency x of the principal gene, the ordinate is the frequency y of the gene that increases mutation. The dotted line is the equilibrium value of x for each y, $\hat{x}$ is the equilibrium of x under selection alone, and x_m is the equilibrium of x under mutation alone. Selection and mutation together move x toward the dotted line while y increases outside the interval $(\hat{x}, x_m)$ and decreases within the interval. (From Levins, 1967.)

average frequency of the lethal is below optimum. Therefore, some mutation toward the lethal would be advantageous. However, this does not prove that the mutation rate would increase. Figure 5.2 explores the direction of selection. For the principle locus, let $\hat{x}$ be the equilibrium value for a given environment under selection alone, and let x_m be the equilibrium value under mutation alone. It can readily be shown that, when x lies between $\hat{x}$ and x_m, an allele increasing mutation rate will be selected against, while if x

lies outside this interval the mutation rate gene will be favored. It can further be seen that, in a constant environment, x always ends up at an equilibrium between $\hat{x}$ and x_m. Only if there is enough fluctuation to keep x outside this interval most of the time can we expect mutation rate to be increased. But it is not yet clear how to define the conditions more precisely, since the extent of selection for mutation depends on linkage. Similar problems arise for genes affecting linkage.

If we approach the genetic system from the point of view of evolution and ecology, the complexities of multiple systems with epistasis can be reduced to relatively few sufficient parameters: (1) memory, (2) delay, (3) ridginess, and (4) multiplicity of peaks.

The emphasis on genetic studies has usually been on equilibrium situations. Experiments with laboratory populations, computer simulation, and the analytical study of simple situations support the view that populations approach the neighborhood of equilibrium rather rapidly (say in 10–30 generations). However, there are other circumstances in which the approach is greatly slowed down. In Figure 5.3 we show the fitness/gene frequency curve for a single locus with simple heterosis. The rate of change depends on $x(1 - x)$ and the slope of the fitness curve. This can be flattened when fitness depends on gene frequency.

Frequency-dependent selection has scarcely been studied systematically, partly because we don't know what models may be biologically meaningful. However, Landahl (unpublished) has considered a situation that may be of general interest. In addition to ordinary selection acting on zygotic

FIGURE 5.3. Fitness as a function of gene frequency for ordinary heterosis.

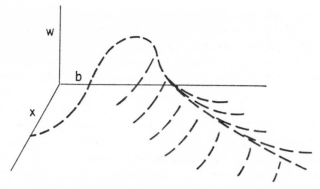

FIGURE 5.4. An adaptive surface where a ridge would show selection.

genotypes, there is an effect of the maternal genotype which acts by determining the sites at which eggs are laid. He found that, when the selective values of the genotypes for choice of egg site and for zygotic viability act in opposite directions, selection can be slowed down greatly, and gene frequencies can be quasi-stable for hundreds of generations.

Multiple-locus situations also permit quasi-equilibrium. Lewontin (1964) found this to happen in simulation studies where there was selection for some intermediate optimum. Here it was the result of selection eliminating the available variance in phenotypic effect rapidly (through elimination of gametic types) while tight linkage prevented fixation of individual loci.

A third type of quasi-equilibrium has been described by William Bossert (1967). Consider an adaptive landscape for two loci, two alleles at each. Suppose that there is a major peak which tapers out into a long ridge as in Figure 5.4.

A population which has reached the lower levels of the ridge should move upward along it. But any random perturbation due to environmental fluctuation, sampling in small populations, or occasional migration will displace the population from the ridge. Much of the response to selection will be expended, returning to or keeping on the ridge rather than moving along it. This effect would increase as we add loci and thus increase the number of dimensions. At eight or nine dimensions the time needed to get near equilibrium may become effectively infinite.

It is clear that the degree of ridginess in the adaptive landscape is a major sufficient parameter of the genetic system. It would be defined mathematically as the ratio of the largest to the smallest eigenvalue of the set of differential equations describing the simultaneous changes of gene frequency. Biologically it will depend on the system of epistatic (non-additive) interactions among loci.

Besides its importance for the study of quasi-equilibrium, the ridginess of an adaptive landscape is one of the points of contact with developmental biology. A developing system depends on genetic interactions. But if these give rise to an adaptive landscape which is too ridgy, selection could not establish that system. This may impose restrictions on the kinds of complexity that can arise in evolution.

Although variable environments have been studied, these have usually been of the simplest kind—pure white noise (environments of successive periods independent) or simple autocorrelation. But there may be more complicated environmental patterns, with combinations of short-term and long-term fluctuations. However, it would be futile to attempt a comprehensive study of possible patterns of environment without more knowledge of the actual patterns of environment.

If in fact we have identified the relevant parameters, the approach of this chapter can be applied as follows:

1. These parameters become objects of study, measurement, and comparison in the description of different groups.

2. A knowledge of the parameters of the genetic system leads to predictions as to the kinds of adaptation which will be possible. For instance, if our argument about epistasis and ridginess is correct, the amount of epistatic interaction should be greater in species which are mostly parthenogentic.

3. Insofar as the genetic system is itself evolving, we should be able to relate the direction of its evolution to the

statistical pattern of the environment. For instance, mutation rates should be higher in species with variable environments and for those loci whose adaptive values fluctuate with the environment.

4. Laboratory experiments could attempt to produce different kinds of changes in the genetic system by natural selection, using only the pattern of environmental change.

5. Theoretical work is needed on the relation of chromosome number, tightness of linkage, and similar genetic parameters to delay, memory, ridginess, and multiple equilibria.

6. The possibilities and limits on second-order selection must be explored systematically.

CHAPTER SIX

From Micro- to Macro-Evolution

Population genetics has developed a model of micro-evolutionary changes which is supported by mathematical theory and observational and laboratory data. It has been shown that selection can produce a response in almost any trait. This response may last only a short time and peter out under inbreeding, or it may persist until a plateau is reached in 20 to 100 generations. This plateau does not indicate that the available genetic variance has been used up, but rather that counterselection is resisting further progress. This counterselection is derived from the integration of the genotype—the dependence of most major traits on many genes and the effect of most genes on many traits. When the frequencies of some of these genes are rapidly displaced from their optimal values for viability and fecundity, the selection intensity pulling them back increases. This is one aspect of the genetic homeostasis discussed by Lerner (1954).

But if a population is held at the plateau for a while, the new imposed frequencies for some of the genes determine the genetic background against which selection occurs for the others. As the residual genotype is reorganized to integrate once more with the altered part, a new cycle of response to selection becomes possible.

Too few studies of this kind have been carried out—and these mostly in *Drosophila*—to permit any comparative analysis of plateauing. However, the level of phenotypic change at which it occurs is a measure of integration of the developmental processes and would be expected to vary systematically among groups. In any case, the plateau marks the upper boundary of the simplest kind of micro-evolution which is best understood. Therefore, the differences between populations can be measured against the difference attained by selection of one population up to the plateau.

The same micro-evolutionary experiments have also been done on divergence of phenotype and the origin of sexual isolation in the laboratory. It has therefore been widely assumed that macro-evolution is simply micro-evolution with more time. However, the theory does not explain why evolution in the paleontological record proceeds perhaps 1/4 million times slower than in laboratory experiments. It does not explain why the rates of evolution vary so much among groups and among periods. It does not explain how an occasional breakthrough into a new niche or new mode of adaption takes place.

The theory outlined in this book added to the previous models an emphasis on the structure of the environment and the notion of adaptive strategy. It can account for differences among taxa in levels of polymorphism, niche breadth, spatial patterns, and, to some extent, the likelihood of speciation.

However, all optimization theory begins by holding something constant and maximizing something else subject to this restriction. The restriction was often expressed in the terms of the fitness set relating fitness to phenotype in several environments simultaneously. But instead of a species optimizing its strategy on a fixed set, it might change the fitness set. In nature, unlike game theory, the plays of the game are not permanently distinct from a change of the rules. Suppose, for example, that thermal adaptation were a function only of the amount of fur. Then fitness at high and low temperature would show a strongly negative correlation and the fitness set would be rather flat, or even concave. But when we add body posture, shivering, or microhabitat selection then traits which help survival in cold do not reduce survival in the heat very much, and the fitness set becomes more convex. If heat and cold adaptation became completely independent, the fitness set would become a rectangle.

The inference is usually made that such a change would constitute a general advance as against the specific adaptation that occurs when an optimum phenotype is calibrated on a fixed fitness set. It is further assumed that a general advance is more likely to occur in "generalized" than in "specialized" species. We will consider both of these questions with respect to change of niche and change in the fitness set.

We have already seen (Figure 2.3a) that on a convex fitness set the optimum strategy is generally intermediate between the phenotypes specialized to one or the other environment. As the proportions of environments change, so does the optimum. If we have a continuum of environments or discrete habitats that are similar enough so that successive habitats along the environmental axis give convex fitness sets, the species can evolve indefinitely into new environments as their proportions change. This holds both for the purely phenotypic fitness set model and for one that specifies genotypes. Therefore it is a robust result.

If there is a concave fitness set, we have to distinguish cases. Fine-grained selection on a concave fitness set results in specialization. Therefore, whenever a gap occurs in available environments which is great enough compared to the tolerance of the individual to give a concave fitness set, the spread of the species will cease. In terms of genetics, each block of environments separated by a concavity constitutes an adaptive peak. The greater the tolerance of the species, the fewer the distinct adaptive peaks and the greater the gap it can span.

When we refer to a gap, we do not require the complete absence of intermediate environments. In Figure 6.1, we show how to calculate the maximum abundance of an intermediate habitat which is effectively a gap.

Course-grained selection has a slightly different effect. In the models that specify genotype, a concave fitness set can still permit a mixed strategy polymorphism provided the concavity is not too extreme (Levins and MacArthur, 1966). The purely phenotypic fitness set model indicates a mixed strategy of specialized types. But if these specialized phenotypes are too different, the new ones could not arise from selection of existing types. Thus the coarse-grained

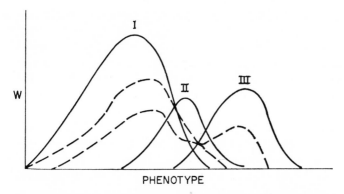

FIGURE 6.1. The concavity gap. The solid curves are fitnesses in single environments. The upper broken curve is the average of fitness in I and II, and the lower curve averages this curve and III. The area under each curve is proportional to its abundance.

pattern (environmental uncertainty) would allow a population to spread across small concavities but not large ones.

None of these models allowed for the effects of population density. Suppose that an organism produces an average of N zygotes. When the population reaches saturation in a constant environment, the average viability of each is $1/N$. In an adjacent environment which the species does not use, the viability of a zygote in the absence of density effects is v. Then if $v > 1/N$ there is a selective advantage to invading the new habitat. The greater the value of N, the greater the environmental gap that can be spanned.

The same type of argument applies to secular changes in in the environment. If the rate of change is slow enough so that the actual population mean is within a "convex" distance of the optimum, the population can follow the environment. Otherwise it will be stranded on an adaptive peak that sinks.

Thus a number of considerations converge to support the objective reality of faunal dominance. A dominant species will be broad-niched, have a broad tolerance, and therefore could expand into new niches across environmental gaps and survive moderately abrupt changes of environment.

101

At first sight there may not seem to be any reason why a species which has a broad range of environmental tolerances should be more likely to tolerate a new kind of environmental stress than would a narrow-niched species. However, as a strange stimulus penetrates the organism it acts by altering the existing systems, changing the existing pathways in directions that are less unique than the external stimulus. For instance, part of the cause of heat death is anoxia. A species may resist high temperature by developing some tolerance to anoxia. It would then also be tolerant to anoxia caused by specific poisons of the cytochrome system which it has never met before.

We now turn to the evolution of the homeostatic system. Consider a fitness set in which the axes are fitness in a constant environment and fitness in a variable environment with the same mean. Since survival in a variable environment involves a homeostatic system which imposes a cost, this will reduce fitness in the constant environment. This convexity or concavity of the fitness set would seem to depend on whether the homeostatic system improves at an accelerating or at a decreasing rate as a function of cost. A fine-grained situation would be one in which the environment changes often within a habitat. As the frequency of change increases, survival under stress becomes more frequent. A coarse-grained pattern would be one in which some microhabitats are more stable, others more variable.

In all of our models, the amount of homeostasis will increase with the amount of environmental variability, although this may take place gradually or abruptly. Each increase in homeostasis makes the fitness set whose axes are different environments more convex. When a threshold is reached, the population can cross an environmental gap. In the concave models and in the convex model with genotypes specified, there is a threshold variability in the environment before increased homeostasis will be selected. Once it does appear, further selection will act to reduce cost so that it will also be advantageous in less variable environments. Thus what appears to hindsight as a general advance, distinct from specific adaptation, can occur as specific adaptation to an environment which is sufficiently variable.

Another type of evolutionary change which is often regarded as general advance is the increase in complexity of structure or activity. This can be interpreted as the evolution of each part in the direction of optimizing some function in the presence of an environment provided by the other parts. Let us look, at the internal organization of the organism. Let "phenotype" refer to alternative enzymes which can carry out the same reaction. The environments may refer to external environmental fluctuations, as in temperature, or internal states, different tissues, or phases of a biological rhythm; the problem then becomes, which enzyme or combination of enzymes is optimal. There are some difficulties involved in defining the Adaptive Function here. If the different environments are interspersed in a very fine-grained way with the enzyme product entering a common pool, or if they suceed each other rapidly in time, the optimum system may be that which maximizes total output and therefore be linear. But if the product of the enzymatic reaction must function in the local site or time of production, and if fitness depends on the joint effects of fitness of each type of cell, a multiplicative Adaptive Function would seem more reasonable. In this line of investigation, the choice of Adaptive Function must be developed along with the results of its use. In any case, the main result is now the following.

If there is a convex fitness set (the different enzymes not too sensitive to the different conditions), the optimal is a single form of enzyme which does moderately well in each condition. But if the heterogeneity of conditions exceeds the range over which each enzyme functions well the fitness set will be concave. Then in a situation where the Adaptive Function is linear, the optimum enzyme is one which has its peak performance in one of the sites or phases of the organism, and may be virtually inactive in other parts. It may or may not be possible for the enzyme to be concentrated at such sites or times. In any case, the specialization of enzyme function increases the heterogeneity of the organism. If the Adaptive Function is more multiplicative, a concave fitness set would give rise to a mixed strategy optimum—a set of isoenzymes with overlapping rate/pH curves.

This may be attained genetically by heterozygosity. Of course, given these isoenzymes, fitness may be further increased by limiting each kind of enzyme to the tissue or stage it is best in. Once more, the adoption of a mixed strategy to meet the internal heterogeneity diversity of the organism itself gives rise to increased heterogeneity.

We can now look at the inverse of this problem. Consider the conditions which defined the original heterogeneity, such as pH. Let this be the phenotype. Now consider that fitness is a multiplicative function of "fitness" with respect to different biological processes such as protein synthesis and sugar utilization. The axes are now the rates of these processes, and the phenotype is pH. On a convex fitness set the optimum pH is intermediate between the optima for the two processes, but on a concave set the result is a mixed strategy—the formation of spatial heterogeneity with respect to pH, or an internal fluctuation of pH to allow both processes to function near optimum part of time.

A similar approach can be taken to interpret the topology of biochemical pathways, the location of loops of various kinds, etc. Here the Adaptive Function is not known, but we can consider several alternatives. For example, suppose that the optimum is either one which minimizes random fluctuations in rate or one which maintains a constant amount of a substance in the face of random changes in the first substrate. In the first optimization, end-product feedback inhibition loops all increase fitness. In the second optimization, loops within loops add nothing to stability. Thus the observations will help determine the Adaptive Function, and the theory as a whole can be used to help explain the evolution of cellular organization. Also, by predicting the types of networks it can direct the biochemical searches in a less haphazard way than the almost random screening of substances.

Morowitz et al. (1964) have taken another approach. They look at the connectedness of the biochemical processes in a global way (that is, without looking at the loops and branches) and conclude that a high degree of connectedness reduces the efficiency of the system but increases the

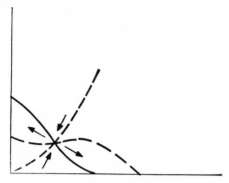

FIGURE 6.2. Coevolution of two subsystems x and y. The solid line is the optimum x for a given y. The broken line is the optimum y for a given x. The arrows show the direction of coevolution and the dotted line divides the plane into two regions separating initial values of x,y into those which lead to one or the other final state.

stability. Therefore, the amount of connectedness which evolves optimally will depend both on the uncertainty of the internal environment, in the sense of perturbations of stability whether of external or internal origin, and on the cost to the system of being perturbed (its tolerance). In any situation in which system A is functioning in the face of fluctuating conditions, there are available the alternatives of stabilizing A by a homeostatic system B, of increasing the tolerance of A at the expense of efficiency. For each degree of tolerance of A there is an optimal B, and for each B an optimum A. At present we have no way of knowing whether the joint evolution of A and B will reach a unique equilibrium or have several alternative equilibria depending on the initial conditions. In our discussion of the joint selection of habitat preference and tolerance we saw examples of both.

Where there are multiple alternatives, the one which is reached depends on the starting point but also on the rates of change in each direction. This is illustrated in Figure 6.2 in which we show the coevolution of two subsystems of an organism. The curves which show the optima are clearly insufficient to predict the course of change alone. The equation for rate of change of each is needed but is likely to be unavailable.

However, the change will clearly be greatest in the direction where the available genetic variance is greatest and where there is least counterselection pressure. Dickerson (1955) showed that where natural selection is acting on N different criteria of fitness equally, an average negative correlation of $-1/N$ between the fitness effects of a given trait over all criteria is enough to block the progress. Therefore, a given parameter which is tightly bound up to other traits will not change readily.

Suppose there is a set of k parameters x_1, x_2, . . . , x_k of phenotype. The functioning of the organism depends on a smaller number of sufficient parameters $F_j(x_1, x_2, . . . , x_k)$, $j = 1, 2 \cdot \cdot \cdot m$, each of which has some optimum value. The fact that $m < k$ means that there are available degrees of freedom for maneuver, and that new selection criteria can also be imposed. It also accounts for the predictability of response at the higher level (e.g., that a population will respond to DDT by the building up of resistance) and the indeterminacy of the detailed response—the distribution of the genes over the chromosomes, the correlated responses, etc.

So far, we have merely restated more abstractly what we already know: That organisms are different, that parts are interconnected, that selection occurs on the basis of what is available. But this frame of reference leads to two complementary principles:

1. *Developmental drift.* If a phyletic line passes through many environments on a very long time scale, the parameters F_j will vary in importance. And for long periods some of the F_j may vanish. This increases the degrees of freedom of the phenotypic variables x_j. When a new selection pressure is imposed, the response will be greatest for the x_i which are most loosely bound, thus establishing a new binding.

The new binding can be visualized as arising in the following way. Consider a fitness set in which the axes are $F_{m+1}(x_1, x_2, . . . , x_k)$, a new sufficient parameter, and

$F^*(x_1, x_2, \ldots, x_k)$ where F^* combines the fitnesses of all F_j up to F_m. Consider optimization to take place by changing x_1 only, the other x_j remaining fixed.

If the fitness set is convex, the optimum value of x will be a compromise between the values maximizing F^* and F_{m+1}. The greater the relative importance of F_m, the more x_1 is determined by that function. The more tightly x is bound to other F_j, the greater the relative weight of F^* compared to F_{m+1} and the less it will change.

If the fitness set is concave, there are two alternatives. With a linear Adaptive Function there is a threshold value for the relative weight of F_{m+1} to have any influence. For this relative weight $P < P_0$, x is determined completely by F^*, and the new constraint F_{m+1} does not alter development. When $P > P_0$, there is a sudden shift in x_1 to the value which is optimal for F_{m+1}. Thus the variable x_1 has been cut loose from determination by F^* to be related only to F_{m+1}. This will then result in modifications of the other x_j to compensate for the shift in x_1.

A multiplicative Adaptive Function results in a mixed strategy, which can be interpreted to mean that x_1 takes on different values in different places or at different times. In effect, x_1 has been split into x_1', a new variable.

In either case, a convex fitness set results in each F_j determined by all the x_j, resulting in a tightly integrated system which will respond to new pressures by gradual adjustment of all the variables. A concave set results in the specialization of variables x_j each to one fuction F_j, so that each F is determined by a relatively few variables which are autonomous. Changes in the selection pressures for a given F results in changes only in these variables.

This model helps account for the origin of degrees of integration and autonomy, and also shows why developmental systems may not respond to new selection pressures even when there is so much genetic variability in populations.

The set of x's which are least tightly bound, which are free to vary, does not depend on the new selection pressure. It will be random with respect to the new pressure although dependent on past history. This random variation is designated "developmental drift." In particular, we propose that those phyletic lines which have passed through many different kinds of environments on a geological time scale will be constantly binding and unbinding variables, will have more loosely bound x_i and more degrees of freedom, and will therefore be more able to undergo major restructuring.

2. *Progressive binding.* Under constant macro-conditions, there will be a tendency toward increased integration among functions that form a convex fitness set. This can come about through selection for cost reduction by making the same variable do multiple duty. Further, the fact that a given character is more or less constant makes it available as a signal to initiate other processes. Thus what is there will become necessary in other ways. The possibility arises that after a very long time in the same environment an organism can become over-integrated in the sense that a given change which may be advantageous in one respect may require too many concomitant changes to be feasible.

Consider, for example, the high-energy phosphate bond which is universally required in the energy utilization of cells. Suppose that in a phosphorus-poor environment an alternative such as a high-energy arsenate might be more available (even though less efficient). However, the ATP intervenes at so many vital points in the metabolism that its replacement by some alternative may require more simultaneous changes than can be expected to arise at once.

The discussion so far leads to two alternative modes of explanation of the differences in evolutionary patterns among groups. The first, the pure environmentalist view, asserts that almost any group can adapt to almost any environment if the selection pattern is appropriate. It assumes that there is virtually infinite genetic variability in populations even for traits that define the higher taxa, but that this variability remains latent. The second approach is developmentalist. This approach insists that there are fundamental differences in the tightness of binding of developmental systems so that, given the same environmental change, some groups simply cannot adapt in time, while others will. Of

108

course the differences in developmental systems also evolved in relation to environment, but the point is that they may be the effective direct causes of evolutionary differences.

Together, these approaches suggest the possibility and the need for experimental work in macro-evolution. Several lines of work seem most promising.

1. *The measurement of the tightness of binding.* The tightness of binding of one phenotypic trait to another is observable as a correlated response to selection and as joint variation in the face of internal or external environmental perturbation. The tight binding of a given trait to the rest of the organism as a whole implies that there are many unidentified selection pressures acting on the genes which also affect a known trait. These pressures will normally be stabilizing and therefore will act as a counterpressure to an imposed selection. Hence the over-all tightness of binding would be measured by

(a) the ratio of the response to selection obtained and that predicted by estimates of heritability;

(b) the response to selection attained before a plateau is reached, as indicated by the height of the plateau and the total selection differential applied to reach it.

(c) The increase in phenotypic variability following the induction of mutation is an inverse measure of binding.

Once we have explored the technical problems in the measurement of binding we can raise the following empirical questions: is over-all binding tighter in some kinds of organisms than in others? If so, is this related to the kind of development, to the kind of environment, to the age or ecological position of the group? Secondly, we can compare the tightness of binding for different traits. Are the phylogenetically older traits more tightly bound? Is the binding tighter for properties of cellular metabolism than for morphological characters? For traits related to the homeostatic system?

Finally, by comparing the correlated responses in different species we can look for evidence of unbinding and rebinding into new clusters of traits.

2. *The mapping of fitness sets.* Since the concavity or convexity of the fitness set played a crucial role in the theory, it is important to know if there are any concave fitness sets in living organisms. The fitness sets would be most readily mapped for biochemical traits, comparing the levels of enzyme activity for different pH or other internal states. Both between organisms and between enzymes the differences in fitness set should be correlated with differences in heterogeneity.

3. *The attempt to produce major evolutionary changes.* It is, of course, very difficult to decide what would be an acceptable criterion of a macro-evolutionary change. However, some of the following components at least are necessary:

(a) The production on the basis of polygenes of a phenotype otherwise not found in the taxon, such as Waddington's bithorax in *Drosophila* which formally transcends the order Diptera.

(b) The successful invasion of a radically new kind of environment.

(c) The functional and morphological divergence of similar structures, such as appendages of a centipede.

(d) Appearance of new mode of adaptation, such as the production of homeothermy in anoline lizards or of a paedogenic *Drosophila*.

Such experiments could make use of the ordinary selection techniques which have been developed, plus special procedures:

(a) The unmasking of hidden variation by the introduction of alien genes. Sometimes a particular gene, such as scute in *Drosophila*, will do this. At other times crosses between almost incompatible populations would be indicated.

(b) Genetic assimilation. Here an environmental shock, usually semilethal, exceeds the bounds of the homeostatic system and reveals hidden genetic differences. These can then be selected for increased effect until the additional environmental stimulus is no longer needed. For instance, the use of insect hormones to uncouple moulting from sexual maturation would be the first step in producing paedogenesis. Thyroxin might be used in the selection for thermoregulation in lizards.

(c) Release from selection pressures by masking. Wherever there exist alternative pathways in cellular metabolism, the blocking of one pathway diverts activity to the other and releases the first for selection in other directions. For instance, we could attempt to alter the cytochrome system in facultative anaerobes under anaerobic conditions. Or in the absence of alternative pathways an exogenous source of nutrient releases the synthetic pathway so that early stages of synthesis can be linked to completely different networks.

Although experiments in macro-evolution are possible in principle, each individual proposal will involve formidable technical difficulties which could only be solved by those who know the particular organism intimately. But in addition, there is the obstacle of time.

With the possible exception of bacteria, the experiments may take very long. Even allowing that selection under optimally designed conditions may proceed 10,000 or 100,000 times faster than in nature, we would need some 10–20 years with *Drosophila* and at least 100 with *Anolis*. Such long-term experiments would, of course, only be feasible if the design is such that there are many short-term results derivable directly or by using the main stock for other purposes, if the amount of labor per year is low so that the fate of the project does not depend on the shifting tastes and interests of the experimenter, and if there are institutional arrangements which permit the continuation of research projects beyond the duration of individual researchers. Perhaps the next major breakthroughs in the field of experimental macro-evolution will depend on previous or concommitant changes in the sociology of our science.

Bibliography

Bossert, W. 1967. Mathematical challenges to the neo-Darwinian interpretation of evolution. *Wistar Inst. Symp. Monograph No. 5.*

Bradshaw, A. D. 1965. Evolutionary significance of phenotypic plasticity in plants. *Advances in Genetics* 13:115–155. Academic Press, Inc., New York.

Brattstrom, B. H., and P. Lawrence. 1962. The rate of thermal acclimation in anuran amphibians. *Physiol. Zool.* 35:148–156.

Brown, W. L., and E. O. Wilson. 1956. Character displacement. *Syst. Zool.* 5:49–64.

Clarke, B. C. 1966. The evolution of morph-ratio clines. *Am. Nat.* 100:389–402.

Clausen, J., D. Keck, and W. M. Heisey. 1940. Experimental studies on the nature of species, I. Carnegie Inst. Wash. publ. 520.

Cody, M. 1966. A general theory of clutch size. *Evol.* 20:174–184.

Cohen, D. 1967a. Optimization of seasonal migratory behavior. *Am. Nat.* 101:5–18.

Cohen, D. 1967b. Optimizing reproduction in a randomly varying environment. *J. Theor. Biol.* 16(1):1–14.

Dickerson, G. E. 1955. Genetic slippage in response to selection for multiple objectives. *Cold Spring Harbor Symp. Quant. Biol.* 20:213–224.

Dobzhansky, T., and C. Pavan. 1950. Local and seasonal variations in relative frequencies of *Drosophila* in Brazil. *J. Anim. Ecol.* 19:1–14.

Elton, Charles. 1927. *Animal Ecology.* Sidgwick & Jackson Ltd., London.

Gause, G. F. 1934. *The Struggle for Existence.* Williams & Wilkins, Baltimore.

Gregor, J. W., and P. Watson. 1961. Ecotype differentiation: observations and reflections. *Evol.* 15(2):166–173.

Grinnell, J. 1904. The origin and distribution of the chestnut-backed chickadee. *Auk.* 21:364–382.

Harrison, J. L. 1962. Distribution of feeding habits among animals in a tropical forest. *J. Anim. Ecol.* 31(1):53–63.

Hutchinson, G. E. 1965. *The Ecological Theatre and the Evolutionary Play.* Yale University Press, New Haven.

Kerner, E. 1957. A statistical mechanics of interacting biological species. *Bull. Math. Biophys.* 19:121–146.

Kimura, M. 1964. *Diffusion Models in Population Genetics.* Metheun Review Series in Applied Probability, Vol. 2.

King, J. 1965. Genetic implications in the origin of higher levels of organization. *Syst. Zool.* 41(4):249–258.

Kurtén, B. 1959. Rates of evolution in fossil mammals. *Cold Spring Harbor Symp. Quant. Biol.* 24:205–216.

Landahl, H. Unpublished manuscript.

Leigh, E. 1965. On the relation between the productivity, biomass, diversity, and stability of a community. *PNAS* 53:777–783.

Lerner, M. 1954. *Genetic Homeostasis.* Oliver & Boyd, Edinburgh.

Levene, H. 1953. Genetic equilibrium when more than one niche is available. *Am. Nat.* 87:331–333.

Levins, R. 1962. Theory of fitness in a heterozeneous environment, I. The fitness set and adaptive function. *Am. Nat.* 96:361–378.

——. 1965. . . . , V. Optimal genetic systems. *Genetics* 52:891–904.

——. 1966. Strategy of model building in population biology. *Am. Sci.* 54:421–431.

——. 1967. . . . , VI. The adaptive significance of mutation. *Genetics* 56:163–178.

Levins R., and R. MacArthur. 1966. Maintenance of genetic polymorphism in a heterogeneous environment: variations on a theme by Howard Levene. *Am. Nat.* 100:585–590.

Lewontin, R. 1964. The interaction of selection and linkage. II. Optimum models. *Genetics* 50:757–782.

——. 1966. Is nature probable or capricious? *Bioscience* 16:25–27.

Lloyd, M., and H. Dybas. 1966. The periodical cicada problem, II. evolution. *Evol.* 20:466–505.

MacArthur, R. 1958. Population ecology of some warblers of northern coniferous forests. *Ecol.* 39:599–619.

MacArthur, R., and E. Pianka. 1966. On the optimal use of a patchy environment. *Am. Nat.* 100:603–609.

Martinez Pico, M., C. Maldonado, and R. Levins. 1965. Ecology and genetics of Puerto Rican *Drosophila*, I. Food preferences of sympatric species. *Carib. J. Sci.* 5:29–38.

Martinez Viera, R. 1964. *Poeyana*, #5.

Morowitz, H. J., W. A. Higinbotham, S. W. Matthysse, and H. Quastler. 1964. Passive stability in metabolic networks. *J. Theor. Biol.* 7:98–111.

Patterson, J. T. 1943. The Drosophilidae of the southwest. University of Texas publ. 4313.

Preston, F. W. 1962. The canonical distribution of commonness and rarity, II. *Ecol.* 43:410–432.

Sacher, G. 1966. The complementarity between development and aging—experimental and theoretical considerations. *N.Y. Acad. Sci. Conf. on Interdisciplinary Perspectives of Time*, Jan. 17–20, 1966.

Schmalhausen, I. I. 1949. *Factors of Evolution*. Blakiston. Philadelphia.

Schoener, T. W. 1965. The evolution of bill size differences among sympatric congeneric species of birds. *Evol.* 19: 189–213.

Tantawy, A. O. and G. S. Mallah. 1961. Studies on natural populations of *Drosophila*, I. heat resistance and geographic variation in *Drosophila melanogaster* and *D. simulans*. *Evol.* 15:1–14.

Vernberg, F. J., and R. E. Tashian. 1959. Studies on the physiological variation between tropical and temperate zone fiddler crabs of the genus Vca. II. Oxygen consumption of whole organisms. *Biol. Bull.* 117:163–184.

Waddington, C. H. 1957. *The Strategy of the Genes*. MacMillan, New York.

Wallace, B., and M. Vetukhiv. 1955. Adaptive organization of the gene pools of Drosophila populations. *Cold Spring Harbor Symp. Quant. Biol.* 20:303–310.

Wilson, E. O. The ergonomics of caste in social insects (unpublished).

Wilson, E. O., and R. W. Taylor. 1967. The estimate of potential evolutionary increase in species density in the Polynesian ant fauna. *Evol.* 21:1–10.

Index

adaptation, 10–13; specific, 99
Adaptive Function, 17, 18, 22,
    23, 26, 27, 31, 32, 69, 70, 74,
    103, 104, 107
advance, general, 99
Agrostis tenuis, 66
annuals, 67
Anolis, 111
ants, 23
aphids, 23
assimilation, genetic, 110
autocorrelation, 35
autonomous development, 21, 22

Beta galactosidase, 10
birds, 39, 48
Bombyx mora, 13
Bossert, William, 95
Bradshaw, A., 38
Brattstrom, B. H., and
    P. Lawrence, 23
breadth of niche, 39, 41, 43, 44,
    46, 53, 99
Brown, W., and E. O. Wilson, 58

canalization, 21
carrying capacity, 5
caste, 23, 24, 37
centipede, 110
Cepaea, 80
character displacement, 58
cicada, 29
Clarke, B., 75, 80
Clausen, J., D. Keck, and
    W. M. Heisey, 66
climax species, niches of, 44
cline, 70
clines, 66, 72, 88
clutch, 25
coarse-grained environment, 74,
    76, 78, 85, 88, 102
Cody, M., 25
coevolution, 58
coexistence, 39, 57
Cohen, D., 28

coherence, genetic, 82, 83, 84,
    88
colonizing species, 44
competition, 50; coefficient of,
    51–53, 58
competitive ability, 25
complexities of nature, 3
complexity, increase in, 6; in
    evolution, 96; of structure, 103
concavity, 17, 19, 32, 69, 85,
    103, 107
connectedness, 104, 105
convergence, 60
convexity, 17, 19, 23, 45, 69
cost of strategy, 34
counterselection, 98
courtship, 80
cross correlation, 35

deciduous forests, 36
delay, parameter of, 94
desiccation, 68
development, dependent, 92
developmental switch, 22
diapause, 12, 13, 23, 26, 37
Dickerson, G. E., 106
diffusion process, 83
dimension of niche, 41, 47
dimensionality, 48, 57
discontinuities, 66, 88
diversity, 39, 41, 49, 87
Dobzhansky, T., and C. Pavan,
    49
dominance, 34; faunal, 101
dormancy, 23, 26
drift, developmental, 106;
    random, 89
Drosophila, 23, 38, 40, 41, 43,
    44, 48, 49, 53, 57, 61, 67, 73,
    87, 88, 98, 110, 111

ecotypes, 66
Elton, C., 39
empiricism, 4
empiricist, 3